Lecture Notes in Networks and Systems   **986**

The series "Lecture Notes in Networks and Systems" publishes the latest developments in Networks and Systems—quickly, informally and with high quality. Original research reported in proceedings and post-proceedings represents the core of LNNS.

Volumes published in LNNS embrace all aspects and subfields of, as well as new challenges in, Networks and Systems.

The series contains proceedings and edited volumes in systems and networks, spanning the areas of Cyber-Physical Systems, Autonomous Systems, Sensor Networks, Control Systems, Energy Systems, Automotive Systems, Biological Systems, Vehicular Networking and Connected Vehicles, Aerospace Systems, Automation, Manufacturing, Smart Grids, Nonlinear Systems, Power Systems, Robotics, Social Systems, Economic Systems and other. Of particular value to both the contributors and the readership are the short publication timeframe and the worldwide distribution and exposure which enable both a wide and rapid dissemination of research output.

The series covers the theory, applications, and perspectives on the state of the art and future developments relevant to systems and networks, decision making, control, complex processes and related areas, as embedded in the fields of interdisciplinary and applied sciences, engineering, computer science, physics, economics, social, and life sciences, as well as the paradigms and methodologies behind them.

Indexed by SCOPUS, INSPEC, WTI Frankfurt eG, zbMATH, SCImago.

All books published in the series are submitted for consideration in Web of Science.

For proposals from Asia please contact Aninda Bose (aninda.bose@springer.com).

Álvaro Rocha · Hojjat Adeli ·
Gintautas Dzemyda · Fernando Moreira ·
Aneta Poniszewska-Marańda
Editors

Good Practices and New Perspectives in Information Systems and Technologies

WorldCIST 2024, Volume 2

 Springer

Editors
Álvaro Rocha
ISEG
Universidade de Lisboa
Lisbon, Portugal

Hojjat Adeli
College of Engineering
The Ohio State University
Columbus, OH, USA

Gintautas Dzemyda
Institute of Data Science and Digital
Technologies
Vilnius University
Vilnius, Lithuania

Fernando Moreira
DCT
Universidade Portucalense
Porto, Portugal

Aneta Poniszewska-Marańda
Institute of Information Technology
Lodz University of Technology
Łódz, Poland

ISSN 2367-3370   ISSN 2367-3389 (electronic)
Lecture Notes in Networks and Systems
ISBN 978-3-031-60217-7   ISBN 978-3-031-60218-4 (eBook)
https://doi.org/10.1007/978-3-031-60218-4

Preface

This book contains a selection of papers accepted for presentation and discussion at the 2024 World Conference on Information Systems and Technologies (WorldCIST'24). This conference had the scientific support of the Lodz University of Technology, Information and Technology Management Association (ITMA), IEEE Systems, Man, and Cybernetics Society (IEEE SMC), Iberian Association for Information Systems and Technologies (AISTI), and Global Institute for IT Management (GIIM). It took place in Lodz city, Poland, 26–28 March 2024.

The World Conference on Information Systems and Technologies (WorldCIST) is a global forum for researchers and practitioners to present and discuss recent results and innovations, current trends, professional experiences, and challenges of modern Information Systems and Technologies research, technological development, and applications. One of its main aims is to strengthen the drive toward a holistic symbiosis between academy, society, and industry. WorldCIST'24 is built on the successes of: WorldCIST'13 held at Olhão, Algarve, Portugal; WorldCIST'14 held at Funchal, Madeira, Portugal; WorldCIST'15 held at São Miguel, Azores, Portugal; WorldCIST'16 held at Recife, Pernambuco, Brazil; WorldCIST'17 held at Porto Santo, Madeira, Portugal; WorldCIST'18 held at Naples, Italy; WorldCIST'19 held at La Toja, Spain; WorldCIST'20 held at Budva, Montenegro; WorldCIST'21 held at Terceira Island, Portugal; WorldCIST'22 held at Budva, Montenegro; and WorldCIST'23, which took place at Pisa, Italy.

The Program Committee of WorldCIST'24 was composed of a multidisciplinary group of 328 experts and those who are intimately concerned with Information Systems and Technologies. They have had the responsibility for evaluating, in a 'blind review' process, the papers received for each of the main themes proposed for the conference: A) Information and Knowledge Management; B) Organizational Models and Information Systems; C) Software and Systems Modeling; D) Software Systems, Architectures, Applications and Tools; E) Multimedia Systems and Applications; F) Computer Networks, Mobility and Pervasive Systems; G) Intelligent and Decision Support Systems; H) Big Data Analytics and Applications; I) Human-Computer Interaction; J) Ethics, Computers & Security; K) Health Informatics; L) Information Technologies in Education; M) Information Technologies in Radiocommunications; and N) Technologies for Biomedical Applications.

The conference also included workshop sessions taking place in parallel with the conference ones. Workshop sessions covered themes such as: ICT for Auditing & Accounting; Open Learning and Inclusive Education Through Information and Communication Technology; Digital Marketing and Communication, Technologies, and Applications; Advances in Deep Learning Methods and Evolutionary Computing for Health Care; Data Mining and Machine Learning in Smart Cities: The role of the technologies in the research of the migrations; Artificial Intelligence Models and Artifacts for Business Intelligence Applications; AI in Education; Environmental data analytics; Forest-Inspired

Computational Intelligence Methods and Applications; Railway Operations, Modeling and Safety; Technology Management in the Electrical Generation Industry: Capacity Building through Knowledge, Resources and Networks; Data Privacy and Protection in Modern Technologies; Strategies and Challenges in Modern NLP: From Argumentation to Ethical Deployment; and Enabling Software Engineering Practices Via Last Development Trends.

WorldCIST'24 and its workshops received about 400 contributions from 47 countries around the world. The papers accepted for oral presentation and discussion at the conference are published by Springer (this book) in six volumes and will be submitted for indexing by WoS, Scopus, EI-Compendex, DBLP, and/or Google Scholar, among others. Extended versions of selected best papers will be published in special or regular issues of leading and relevant journals, mainly JCR/SCI/SSCI and Scopus/EI-Compendex indexed journals.

We acknowledge all of those that contributed to the staging of WorldCIST'24 (authors, committees, workshop organizers, and sponsors). We deeply appreciate their involvement and support that was crucial for the success of WorldCIST'24.

March 2024

Álvaro Rocha
Hojjat Adeli
Gintautas Dzemyda
Fernando Moreira
Aneta Poniszewska-Marańda

Organization

Conference

Honorary Chair

Hojjat Adeli   The Ohio State University, USA

General Chair

Álvaro Rocha   ISEG, University of Lisbon, Portugal

Co-chairs

Gintautas Dzemyda   Vilnius University, Lithuania
Sandra Costanzo   University of Calabria, Italy

Workshops Chair

Fernando Moreira   Portucalense University, Portugal

Local Organizing Committee

Bożena Borowska   Lodz University of Technology, Poland
Łukasz Chomątek   Lodz University of Technology, Poland
Joanna Ochelska-Mierzejewska   Lodz University of Technology, Poland
Aneta Poniszewska-Marańda   Lodz University of Technology, Poland

Advisory Committee

Ana Maria Correia (Chair)   University of Sheffield, UK
Brandon Randolph-Seng   Texas A&M University, USA

Chris Kimble   KEDGE Business School & MRM, UM2,
                                 Montpellier, France
Damian Niwiński   University of Warsaw, Poland
Eugene Spafford   Purdue University, USA
Florin Gheorghe Filip   Romanian Academy, Romania
Janusz Kacprzyk   Polish Academy of Sciences, Poland
João Tavares   University of Porto, Portugal
Jon Hall   The Open University, UK
John MacIntyre   University of Sunderland, UK
Karl Stroetmann   Empirica Communication & Technology
                                 Research, Germany
Marjan Mernik   University of Maribor, Slovenia
Miguel-Angel Sicilia   University of Alcalá, Spain
Mirjana Ivanovic   University of Novi Sad, Serbia
Paulo Novais   University of Minho, Portugal
Sami Habib   Kuwait University, Kuwait
Wim Van Grembergen   University of Antwerp, Belgium

Program Committee Co-chairs

Adam Wojciechowski   Lodz University of Technology, Poland
Aneta Poniszewska-Marańda   Lodz University of Technology, Poland

Program Committee

Abderrahmane Ez-zahout   Mohammed V University, Morocco
Adriana Peña Pérez Negrón   Universidad de Guadalajara, Mexico
Adriani Besimi   South East European University, North
                                 Macedonia
Agostinho Sousa Pinto   Polytechnic of Porto, Portugal
Ahmed El Oualkadi   Abdelmalek Essaadi University, Morocco
Akex Rabasa   University Miguel Hernandez, Spain
Alanio de Lima   UFC, Brazil
Alba Córdoba-Cabús   University of Malaga, Spain
Alberto Freitas   FMUP, University of Porto, Portugal
Aleksandra Labus   University of Belgrade, Serbia
Alessio De Santo   HE-ARC, Switzerland
Alexandru Vulpe   University Politechnica of Bucharest, Romania
Ali Idri   ENSIAS, University Mohamed V, Morocco
Alicia García-Holgado   University of Salamanca, Spain

Christos Chrysoulas	London South Bank University, UK
Christos Chrysoulas	Edinburgh Napier University, UK
Ciro Martins	University of Aveiro, Portugal
Claudio Sapateiro	Polytechnic of Setúbal, Portugal
Cosmin Striletchi	Technical University of Cluj-Napoca, Romania
Costin Badica	University of Craiova, Romania
Cristian García Bauza	PLADEMA-UNICEN-CONICET, Argentina
Cristina Caridade	Polytechnic of Coimbra, Portugal
Danish Jamil	Malaysia University of Science and Technology, Malaysia
David Cortés-Polo	University of Extremadura, Spain
David Kelly	University College London, UK
Daria Bylieva	Peter the Great St. Petersburg Polytechnic University, Russia
Dayana Spagnuelo	Vrije Universiteit Amsterdam, Netherlands
Dhouha Jaziri	University of Sousse, Tunisia
Dmitry Frolov	HSE University, Russia
Dulce Mourato	ISTEC - Higher Advanced Technologies Institute Lisbon, Portugal
Edita Butrime	Lithuanian University of Health Sciences, Lithuania
Edna Dias Canedo	University of Brasilia, Brazil
Egils Ginters	Riga Technical University, Latvia
Ekaterina Isaeva	Perm State University, Russia
Eliana Leite	University of Minho, Portugal
Enrique Pelaez	ESPOL University, Ecuador
Eriks Sneiders	Stockholm University, Sweden; Esteban Castellanos ESPE, Ecuador
Fatima Azzahra Amazal	Ibn Zohr University, Morocco
Fernando Bobillo	University of Zaragoza, Spain
Fernando Molina-Granja	National University of Chimborazo, Ecuador
Fernando Moreira	Portucalense University, Portugal
Fernando Ribeiro	Polytechnic Castelo Branco, Portugal
Filipe Caldeira	Polytechnic of Viseu, Portugal
Filippo Neri	University of Naples, Italy
Firat Bestepe	Republic of Turkey Ministry of Development, Turkey
Francesco Bianconi	Università degli Studi di Perugia, Italy
Francisco García-Peñalvo	University of Salamanca, Spain
Francisco Valverde	Universidad Central del Ecuador, Ecuador
Frederico Branco	University of Trás-os-Montes e Alto Douro, Portugal
Galim Vakhitov	Kazan Federal University, Russia

Gayo Diallo   University of Bordeaux, France
Gabriel Pestana   Polytechnic Institute of Setubal, Portugal
Gema Bello-Orgaz   Universidad Politecnica de Madrid, Spain
George Suciu   BEIA Consult International, Romania
Ghani Albaali   Princess Sumaya University for Technology,
                                       Jordan
Gian Piero Zarri   University Paris-Sorbonne, France
Giovanni Buonanno   University of Calabria, Italy
Gonçalo Paiva Dias   University of Aveiro, Portugal
Goreti Marreiros   ISEP/GECAD, Portugal
Habiba Drias   University of Science and Technology Houari
                                       Boumediene, Algeria
Hafed Zarzour   University of Souk Ahras, Algeria
Haji Gul   City University of Science and Information
                                       Technology, Pakistan
Hakima Benali Mellah   Cerist, Algeria
Hamid Alasadi   Basra University, Iraq
Hatem Ben Sta   University of Tunis at El Manar, Tunisia
Hector Fernando Gomez Alvarado   Universidad Tecnica de Ambato, Ecuador
Hector Menendez   King's College London, UK
Hélder Gomes   University of Aveiro, Portugal
Helia Guerra   University of the Azores, Portugal
Henrique da Mota Silveira   University of Campinas (UNICAMP), Brazil
Henrique S. Mamede   University Aberta, Portugal
Henrique Vicente   University of Évora, Portugal
Hicham Gueddah   University Mohammed V in Rabat, Morocco
Hing Kai Chan   University of Nottingham Ningbo China, China
Igor Aguilar Alonso   Universidad Nacional Tecnológica de Lima Sur,
                                       Peru
Inês Domingues   University of Coimbra, Portugal
Isabel Lopes   Polytechnic of Bragança, Portugal
Isabel Pedrosa   Coimbra Business School - ISCAC, Portugal
Isaías Martins   University of Leon, Spain
Issam Moghrabi   Gulf University for Science and Technology,
                                       Kuwait
Ivan Armuelles Voinov   University of Panama, Panama
Ivan Dunđer   University of Zagreb, Croatia
Ivone Amorim   University of Porto, Portugal
Jaime Diaz   University of La Frontera, Chile
Jan Egger   IKIM, Germany
Jan Kubicek   Technical University of Ostrava, Czech Republic
Jeimi Cano   Universidad de los Andes, Colombia

Jesús Gallardo Casero	University of Zaragoza, Spain
Jezreel Mejia	CIMAT, Unidad Zacatecas, Mexico
Jikai Li	The College of New Jersey, USA
Jinzhi Lu	KTH-Royal Institute of Technology, Sweden
Joao Carlos Silva	IPCA, Portugal
João Manuel R. S. Tavares	University of Porto, FEUP, Portugal
João Paulo Pereira	Polytechnic of Bragança, Portugal
João Reis	University of Aveiro, Portugal
João Reis	University of Lisbon, Portugal
João Rodrigues	University of the Algarve, Portugal
João Vidal de Carvalho	Polytechnic of Porto, Portugal
Joaquin Nicolas Ros	University of Murcia, Spain
John W. Castro	University de Atacama, Chile
Jorge Barbosa	Polytechnic of Coimbra, Portugal
Jorge Buele	Technical University of Ambato, Ecuador; Jorge Gomes University of Lisbon, Portugal
Jorge Oliveira e Sá	University of Minho, Portugal
José Braga de Vasconcelos	Universidade Lusófona, Portugal
Jose M. Parente de Oliveira	Aeronautics Institute of Technology, Brazil
José Machado	University of Minho, Portugal
José Paulo Lousado	Polytechnic of Viseu, Portugal
Jose Quiroga	University of Oviedo, Spain
Jose Silvestre Silva	Academia Military, Portugal
Jose Torres	University Fernando Pessoa, Portugal
Juan M. Santos	University of Vigo, Spain
Juan Manuel Carrillo de Gea	University of Murcia, Spain
Juan Pablo Damato	UNCPBA-CONICET, Argentina
Kalinka Kaloyanova	Sofia University, Bulgaria
Kamran Shaukat	The University of Newcastle, Australia
Katerina Zdravkova	University Ss. Cyril and Methodius, North Macedonia
Khawla Tadist	Morocco
Khalid Benali	LORIA - University of Lorraine, France
Khalid Nafil	Mohammed V University in Rabat, Morocco
Korhan Gunel	Adnan Menderes University, Turkey
Krzysztof Wolk	Polish-Japanese Academy of Information Technology, Poland
Kuan Yew Wong	Universiti Teknologi Malaysia (UTM), Malaysia
Kwanghoon Kim	Kyonggi University, South Korea
Laila Cheikhi	Mohammed V University in Rabat, Morocco
Laura Varela-Candamio	Universidade da Coruña, Spain
Laurentiu Boicescu	E.T.T.I. U.P.B., Romania

Lbtissam Abnane	ENSIAS, Morocco
Lia-Anca Hangan	Technical University of Cluj-Napoca, Romania
Ligia Martinez	CECAR, Colombia
Lila Rao-Graham	University of the West Indies, Jamaica
Liliana Ivone Pereira	Polytechnic of Cávado and Ave, Portugal
Łukasz Tomczyk	Pedagogical University of Cracow, Poland
Luis Alvarez Sabucedo	University of Vigo, Spain
Luís Filipe Barbosa	University of Trás-os-Montes e Alto Douro
Luis Mendes Gomes	University of the Azores, Portugal
Luis Pinto Ferreira	Polytechnic of Porto, Portugal
Luis Roseiro	Polytechnic of Coimbra, Portugal
Luis Silva Rodrigues	Polytencic of Porto, Portugal
Mahdieh Zakizadeh	MOP, Iran
Maksim Goman	JKU, Austria
Manal el Bajta	ENSIAS, Morocco
Manuel Antonio Fernández-Villacañas Marín	Technical University of Madrid, Spain
Manuel Ignacio Ayala Chauvin	University Indoamerica, Ecuador
Manuel Silva	Polytechnic of Porto and INESC TEC, Portugal
Manuel Tupia	Pontifical Catholic University of Peru, Peru
Manuel Au-Yong-Oliveira	University of Aveiro, Portugal
Marcelo Mendonça Teixeira	Universidade de Pernambuco, Brazil
Marciele Bernardes	University of Minho, Brazil
Marco Ronchetti	Universita' di Trento, Italy
Mareca María Pilar	Universidad Politécnica de Madrid, Spain
Marek Kvet	Zilinska Univerzita v Ziline, Slovakia
Maria João Ferreira	Universidade Portucalense, Portugal
Maria José Sousa	University of Coimbra, Portugal
María Teresa García-Álvarez	University of A Coruna, Spain
Maria Sokhn	University of Applied Sciences of Western Switzerland, Switzerland
Marijana Despotovic-Zrakic	Faculty Organizational Science, Serbia
Marilio Cardoso	Polytechnic of Porto, Portugal
Mário Antunes	Polytechnic of Leiria & CRACS INESC TEC, Portugal
Marisa Maximiano	Polytechnic Institute of Leiria, Portugal
Marisol Garcia-Valls	Polytechnic University of Valencia, Spain
Maristela Holanda	University of Brasilia, Brazil
Marius Vochin	E.T.T.I. U.P.B., Romania
Martin Henkel	Stockholm University, Sweden
Martín López Nores	University of Vigo, Spain
Martin Zelm	INTEROP-VLab, Belgium

Mazyar Zand	MOP, Iran
Mawloud Mosbah	University 20 Août 1955 of Skikda, Algeria
Michal Adamczak	Poznan School of Logistics, Poland
Michal Kvet	University of Zilina, Slovakia
Miguel Garcia	University of Oviedo, Spain
Mircea Georgescu	Al. I. Cuza University of Iasi, Romania
Mirna Muñoz	Centro de Investigación en Matemáticas A.C., Mexico
Mohamed Hosni	ENSIAS, Morocco
Monica Leba	University of Petrosani, Romania
Nadesda Abbas	UBO, Chile
Narasimha Rao Vajjhala	University of New York Tirana, Tirana
Narjes Benameur	Laboratory of Biophysics and Medical Technologies of Tunis, Tunisia
Natalia Grafeeva	Saint Petersburg University, Russia
Natalia Miloslavskaya	National Research Nuclear University MEPhI, Russia
Naveed Ahmed	University of Sharjah, United Arab Emirates
Neeraj Gupta	KIET group of institutions Ghaziabad, India
Nelson Rocha	University of Aveiro, Portugal
Nikola S. Nikolov	University of Limerick, Ireland
Nicolas de Araujo Moreira	Federal University of Ceara, Brazil
Nikolai Prokopyev	Kazan Federal University, Russia
Niranjan S. K.	JSS Science and Technology University, India
Noemi Emanuela Cazzaniga	Politecnico di Milano, Italy
Noureddine Kerzazi	Polytechnique Montréal, Canada
Nuno Melão	Polytechnic of Viseu, Portugal
Nuno Octávio Fernandes	Polytechnic of Castelo Branco, Portugal
Nuno Pombo	University of Beira Interior, Portugal
Olga Kurasova	Vilnius University, Lithuania
Olimpiu Stoicuta	University of Petrosani, Romania
Patricia Quesado	Polytechnic of Cávado and Ave, Portugal
Patricia Zachman	Universidad Nacional del Chaco Austral, Argentina
Paula Serdeira Azevedo	University of Algarve, Portugal
Paula Dias	Polytechnic of Guarda, Portugal
Paulo Alejandro Quezada Sarmiento	University of the Basque Country, Spain
Paulo Maio	Polytechnic of Porto, ISEP, Portugal
Paulvanna Nayaki Marimuthu	Kuwait University, Kuwait
Paweł Karczmarek	The John Paul II Catholic University of Lublin, Poland

Pedro Rangel Henriques	University of Minho, Portugal
Pedro Sobral	University Fernando Pessoa, Portugal
Pedro Sousa	University of Minho, Portugal
Philipp Jordan	University of Hawaii at Manoa, USA
Piotr Kulczycki	Systems Research Institute, Polish Academy of Sciences, Poland
Prabhat Mahanti	University of New Brunswick, Canada
Rabia Azzi	Bordeaux University, France
Radu-Emil Precup	Politehnica University of Timisoara, Romania
Rafael Caldeirinha	Polytechnic of Leiria, Portugal
Raghuraman Rangarajan	Sequoia AT, Portugal
Radhakrishna Bhat	Manipal Institute of Technology, India
Raiani Ali	Hamad Bin Khalifa University, Qatar
Ramadan Elaiess	University of Benghazi, Libya
Ramayah T.	Universiti Sains Malaysia, Malaysia
Ramazy Mahmoudi	University of Monastir, Tunisia
Ramiro Gonçalves	University of Trás-os-Montes e Alto Douro & INESC TEC, Portugal
Ramon Alcarria	Universidad Politécnica de Madrid, Spain
Ramon Fabregat Gesa	University of Girona, Spain
Ramy Rahimi	Chungnam National University, South Korea
Reiko Hishiyama	Waseda University, Japan
Renata Maria Maracho	Federal University of Minas Gerais, Brazil
Renato Toasa	Israel Technological University, Ecuador
Reyes Juárez Ramírez	Universidad Autonoma de Baja California, Mexico
Rocío González-Sánchez	Rey Juan Carlos University, Spain
Rodrigo Franklin Frogeri	University Center of Minas Gerais South, Brazil
Ruben Pereira	ISCTE, Portugal
Rui Alexandre Castanho	WSB University, Poland
Rui S. Moreira	UFP & INESC TEC & LIACC, Portugal
Rustam Burnashev	Kazan Federal University, Russia
Saeed Salah	Al-Quds University, Palestine
Said Achchab	Mohammed V University in Rabat, Morocco
Sajid Anwar	Institute of Management Sciences Peshawar, Pakistan
Sami Habib	Kuwait University, Kuwait
Samuel Sepulveda	University of La Frontera, Chile
Sara Luis Dias	Polytechnic of Cávado and Ave, Portugal
Sandra Costanzo	University of Calabria, Italy
Sandra Patricia Cano Mazuera	University of San Buenaventura Cali, Colombia
Sassi Sassi	FSJEGJ, Tunisia

Seppo Sirkemaa	University of Turku, Finland
Sergio Correia	Polytechnic of Portalegre, Portugal
Shahnawaz Talpur	Mehran University of Engineering & Technology Jamshoro, Pakistan
Shakti Kundu	Manipal University Jaipur, Rajasthan, India
Shashi Kant Gupta	Eudoxia Research University, USA
Silviu Vert	Politehnica University of Timisoara, Romania
Simona Mirela Riurean	University of Petrosani, Romania
Slawomir Zolkiewski	Silesian University of Technology, Poland
Solange Rito Lima	University of Minho, Portugal
Sonia Morgado	ISCPSI, Portugal
Sonia Sobral	Portucalense University, Portugal
Sorin Zoican	Polytechnic University of Bucharest, Romania
Souraya Hamida	Batna 2 University, Algeria
Stalin Figueroa	University of Alcala, Spain
Sümeyya Ilkin	Kocaeli University, Turkey
Syed Asim Ali	University of Karachi, Pakistan
Syed Nasirin	Universiti Malaysia Sabah, Malaysia
Tatiana Antipova	Institute of Certified Specialists, Russia
Tatianna Rosal	University of Trás-os-Montes e Alto Douro, Portugal
Tero Kokkonen	JAMK University of Applied Sciences, Finland
The Thanh Van	HCMC University of Food Industry, Vietnam
Thomas Weber	EPFL, Switzerland
Timothy Asiedu	TIM Technology Services Ltd., Ghana
Tom Sander	New College of Humanities, Germany
Tomasz Kisielewicz	Warsaw University of Technology
Tomaž Klobučar	Jozef Stefan Institute, Slovenia
Toshihiko Kato	University of Electro-communications, Japan
Tuomo Sipola	Jamk University of Applied Sciences, Finland
Tzung-Pei Hong	National University of Kaohsiung, Taiwan
Valentim Realinho	Polytechnic of Portalegre, Portugal
Valentina Colla	Scuola Superiore Sant'Anna, Italy
Valerio Stallone	ZHAW, Switzerland
Verónica Vasconcelos	Polytechnic of Coimbra, Portugal
Vicenzo Iannino	Scuola Superiore Sant'Anna, Italy
Vitor Gonçalves	Polytechnic of Bragança, Portugal
Victor Alves	University of Minho, Portugal
Victor Georgiev	Kazan Federal University, Russia
Victor Hugo Medina Garcia	Universidad Distrital Francisco José de Caldas, Colombia
Victor Kaptelinin	Umeå University, Sweden

Contents

Intelligent and Decision Support Systems

MAGNAT: Maritime Management Ensemble Learning System 3
Niusha Mesgaribarzi

Stock Market Prediction: Integrating Explainable AI with Conv2D Models
for Candlestick Image Analysis 13
Joao Paulo Euko, Flavio Santos, and Paulo Novais

Resources Optimization and Value-Based Prioritization for at Risk
Cultural Heritage Assets Management 24
Ulysse Rosselet and Cédric Gaspoz

Expert Systems in Information Security: A Comprehensive Exploration
of Awareness Strategies Against Social Engineering Attacks 34
*Waldson Rodrigues Cardoso, Admilson de Ribamar Lima Ribeiro,
and João Marco Cardoso da Silva*

Multi-class Model to Predict Pain on Lower Limb Intermittent
Claudication Patients 44
*Rafael Martins, Luís Conceição, Gustavo Corrente, William Xavier,
Júlio Souza, Alberto Freitas, and Goreti Marreiros*

Collaborative Filtering Recommendation Systems Based on Deep
Learning: An Experimental Study 54
*Eddy Pardo, Priscila Valdiviezo-Diaz, Luis Barba-Guaman,
and Janneth Chicaiza*

Assessment of LSTM and GRU Models to Predict the Electricity
Production from Biogas in a Wastewater Treatment Plant 64
*Pedro Oliveira, Francisco S. Marcondes, M. Salomé Duarte,
Dalila Durães, Gilberto Martins, and Paulo Novais*

Fusing Temporal and Contextual Features for Enhanced Traffic Volume
Prediction 74
*Sara Balderas-Díaz, Gabriel Guerrero-Contreras, Andrés Muñoz,
and Juan Boubeta-Puig*

Target-vs-One and Target-vs-All Classification of Epilepsy Using Deep
Learning Technique   85
    Adnan Amin, Feras Al-Obeidat, Nasir Ahmed Algeelani,
    Ahmed Shudaiber, and Fernando Moreira

Health Informatics

OralDentalSoft: Open-Source Web Application for Dental Office
Management   97
    Ricardo Burbano, Eduardo Estévez, Lucrecia Llerena,
    and Nancy Rodríguez

A Scoping Review of the Use of Blockchain and Machine Learning
in Medical Imaging Applications   107
    João Pavão, Rute Bastardo, and Nelson Pacheco Rocha

The Role of Electronic Health Records to Identify Risk Factors
for Developing Long COVID: A Scoping Review   118
    Ema Santos, Afonso Fernandes, Manuel Graça,
    and Nelson Pacheco Rocha

Machine Learning Approaches to Support Medical Imaging Diagnosis
of Pancreatic Cancer – A Scoping Review   129
    Florbela Tavares, Gilberto Rosa, Inês Henriques,
    and Nelson Pacheco Rocha

Virtual Reality in the Pain Management of Pediatric Burn Patients,
A Scoping Review   139
    Joana Santos, Jorge Marques, João Pacheco, and Nelson Pacheco Rocha

Defining the "Smart Hospital": A Literature Review   150
    Leonidas Anthopoulos, Maria Karakidi, and Dimitrios Tselios

Deep Learning for Healthcare: A Web-Microservices System Ready
for Chest Pathology Detection   158
    Sebastián Quevedo, Hamed Behzadi-Khormouji, Federico Domínguez,
    and Enrique Peláez

Risk Factors in the Implementation of Information Systems in a Federal
University Hospital   170
    Eliane Cunha Marques, Simone B. S. Monteiro, Viviane V. F. Grubisic,
    and Ricardo Matos Chaim

The Challenges of Blockchain in Healthcare Entrepreneurship   188
    Maria José Sousa, Miguel Sousa, and Álvaro Rocha

Determinants Associated with Treatment Discontinuation in Tacna Health
Network Tuberculosis Patients 199
Alex Eduardo Tapia- Tenorio, Kevin Mario Laura-De La Cruz,
Roberto Daniel Ballon-Bahamondes,
Luz Anabella Mendoza-Del Valle, Amanda Hilda Koctong-Choy,
Pedro Ronald Cárdenas-Rueda, and Jose Giancarlo Tozo-Burgos

Comprehensive Analysis of Feature Extraction Methods for Emotion
Recognition on Motor Imagery from Multichannel EEG Recordings 211
Amr F. Mohamed and Vacius Jusas

Control of Respiratory Ventilators Using Boussignac Valve 232
Zbigniew Szkulmowski, Sławomir Grzelak, Michał Joachimiak,
Sebastian Meszyński, Marcin Schiller, and Oleksandr Sokolov

Deep Learning Brain MRI Segmentation and 3D Reconstruction:
Evaluation of Hippocampal Atrophy in Mesial Temporal Lobe Epilepsy 243
Aymen Chaouch, Nada Hadj Messaoud, Asma Ben Abdallah,
Jamal Saad, Laurent Payen, Badii Hmida, and M. Hedi Bedoui

Integrating Explainable AI: Breakthroughs in Medical Diagnosis
and Surgery 254
Ana Henriques, Henrique Parola, Raquel Gonçalves,
and Manuel Rodrigues

Author Index 273

Intelligent and Decision Support Systems

MAGNAT: Maritime Management Ensemble Learning System

Niusha Mesgaribarzi[✉] ⓘ

Faculty of Technology, Natural and Maritime Sciences, University of South-Eastern Norway
(USN), Horten, Norway
niusha@ieee.org

Abstract. The classification of ships using remotely sensed data is increasingly
vital for maritime security, environmental monitoring, and commercial applications. Advances in satellite imagery and remote sensing technologies have heightened the need for effective ship classification algorithms. A key challenge lies in
the variability of ship appearances due to size, shape, orientation, and environmental factors. Ensemble learning, which combines multiple classifiers for improved
accuracy, has shown promise in this area. However, this method faces issues like
increased memory and time complexity, and the influence of lower-quality models,
highlighting the need for model pruning. This study presents a new approach to
maritime management through the introduction of an ensemble learning system.
The MAGNAT (Maritime mAnaGement eNsemble leArning sysTem) proposal
and novel contribution in this paper represent an intelligent ensemble solution
that significantly enhances ship classification in the maritime domain through the
adoption of ensemble learning techniques. Incorporating various Convolutional
Neural Network (CNN) models into the training process, a novel aggregation strategy is devised to eliminate redundant models, retaining only those that actively
enhance the learning process. Extensive experiments utilizing a renowned ship
dataset were conducted to validate the effectiveness of the framework, ultimately
demonstrating its superiority over conventional ensemble solutions.

Keywords: Ensemble Learning · Maritime Applications · Convolution Neural
Network

1 Introduction

The classification of ships from remotely sensed data has gained significant importance in
various safety industrial applications [1–3]. Whether for maritime security, environmental monitoring, or commercial interests, accurate and efficient ship classification plays
a pivotal role. In recent years, with the proliferation of satellite imagery, radar systems,
and other remote sensing technologies, the need for robust ship classification algorithms
has grown substantially [4, 5]. One of the significant challenges in ship classification
is dealing with the variability in ship appearances due to factors such as size, shape,
orientation, and environmental conditions. To address this challenge, the utilization of

Á. Rocha et al. (Eds.): WorldCIST 2024, LNNS 986, pp. 3–12, 2024.
https://doi.org/10.1007/978-3-031-60218-4_1

ensemble learning techniques has emerged as a promising approach. Ensemble learning is a methodology that combines the outputs of multiple base classifiers to obtain a more accurate and robust classification result [6–8]. It leverages the strength of diverse models to mitigate the weaknesses of individual classifiers. In the context of ship classification, ensemble learning methodologies have shown remarkable potential for improving classification accuracy, handling data imbalance, and enhancing the overall reliability of classification systems [9–11]. Even though these ensemble methods outperform single learning models, they have several shortcomings: 1. The memory and time complexity linearly increase with the ensemble size (increasing in the number of models trained). 2. Models of poor quality greatly influence the best models in the ensemble. The need for pruning models that contribute negatively to the learning process is primordial for ship classification.

This paper aims to provide an in-depth exploration of pruning ensemble learning in ship classification. In this context, the author will look at the architecture specification of each model in the ensemble and explore these specifications to prune the redundant models. The research presented in this paper contributes to the growing body of knowledge on ship classification and offers valuable insights for researchers, practitioners, and decision-makers seeking to enhance the accuracy and reliability of ship detection and classification systems in diverse applications. By adopting ensemble learning techniques, improvement in the state-of-the-art in ship classification and advancement of the ability to monitor and manage maritime activities can be achieved. The main contribution of the paper is given as follows:

1. The author proposes MAGNAT (Maritime mAnaGement eNsemble leArning sys-Tem), an intelligent ensemble learning solution for ship classification in maritime setting.
2. The author develops an intelligent aggregation method that only selects the models that contribute positively to the learning process by exploring the specification architecture of each model in the ensemble.
3. The author evaluates MAGNAT using a well-known ship dataset. The results reveal MAGNAT's superiority compared to baseline ensemble learning models.

This paper's structure: Sect. 2 reviews ship classification techniques; Sect. 3 describes the proposed methodology; Sect. 4 evaluates performance, and Sect. 5 discusses implications and future research; the conclusion is in Sect. 6.

2   Related Work

Yan et al. [9] introduced a ship classification technique for satellite images that utilize ensemble learning with multiple classifiers. It primarily involves three key steps: initially, they created a ship object classification dataset by utilizing global space-based AIS data; subsequently, they built an ensemble learning model by combining various base classifiers; and finally, they transfer the trained classification model to predict ship types in satellite images. Wang et al. [10] introduced a Multi-Feature Ensemble Learning Classification Model (MFELCM). It followed a three-step process. Firstly, they preprocessed the original data by extracting both static and dynamic information, resulting in static feature samples, dynamic feature distribution samples, time-series samples, and time-series feature samples. In the second step, they trained four base classifiers, namely Random Forest, 1D-CNN (one-dimensional convolutional neural network), Bi-GRU (bidirectional gated recurrent unit), and XGBoost (extreme gradient boosting), using the four types of samples mentioned earlier. Finally, these base classifiers are combined through another Random Forest, resulting in the final output of ship classification. Mostafa et al. [11] introduced an innovative CNN classification approach tailored for inland waterways, capable of categorizing five prominent ship types: cargo, military, carrier, cruise, and tanker. This method is adaptable for various ship classes as well. It comprised four key phases aimed at enhancing classification accuracy within the realm of Intelligent Transport Systems (ITS) using CNNs. These phases encompassed an efficient augmentation method, the utilization of the hyper-parameter optimization (HPO) technique to fine-tune CNN model parameters, the incorporation of transfer learning, and the application of ensemble learning. Zheng et al. [12] introduced an automated approach for assembling a heterogeneous ensemble of Deep Convolutional Neural Networks (DCNNs) through a two-stage filtration process, which they referred to as "MetaBoost." This method effectively enhanced the robustness and accuracy of SAR ship classification. The core concept behind MetaBoost involved creating a diverse pool of heterogeneous classifiers, choosing a subset comprised of the most diverse and accurate classifiers, and ultimately merging meta-features from this optimal subset. Notably, MetaBoost is a self-configuring algorithm capable of autonomously determining the ideal type and quantity of base classifiers for the fusion process. Gao et al. [13] put forth a ship trim optimization technique that harnesses operational data and ensemble learning to attain energy conservation and emissions reduction for ships navigating inland seas. The study followed a three-step approach: a) Initial data processing and the selection of pertinent features from operational data obtained from an inland ro-ro passenger ship. b) The creation of energy consumption prediction models through ensemble learning techniques. c) The formulation of a trim optimization model is achieved by merging the energy consumption model with the most accurate prediction performance and employing an enumeration method. Wei et al. [14] introduced a time-varying ensemble model, aimed at enhancing real-time ship motion prediction performance. The model comprised three key modules, each enhancing its effectiveness. Module I deployed the NSGA-II algorithm for feature selection from the original multi-factor data, eliminating non-essential features for ship motion prediction. Module II utilized the self-organizing map algorithm to cluster the feature-selected data, organizing similar samples into a few clusters. Subsequently, an ensemble learning model, incorporating multiple Elman neural networks and Adaboost techniques, is established for each cluster. Module III presented a time-varying prediction framework for real-time ship motion prediction by amalgamating the ensemble model associated with each cluster. Liu et al. [15] presented a cascaded detection approach for pinpointing and categorizing ships in high-resolution satellite images. Initially, candidate ship hulls were swiftly identified using the phase spectrum of the Fourier transform. Subsequently, a hull refinement module was employed to enhance the precision of candidate hull shapes. To eliminate false alarms, the shape characteristics and textures of candidate hulls were utilized. The likelihood of a candidate hull being identified as a genuine one was enhanced by the presence of ship wakes. Once authentic ships were identified, ship classification was performed using a fuzzy classifier that incorporated information from both the ship hull and wakes.

Ensemble learning techniques for ship classification entail the consolidation of predictions from multiple independently trained learning models. Despite their demonstrated superiority over individual models, these methods exhibit certain limitations. Specifically: 1. The memory and computational time complexity exhibit linear growth with the size of the ensemble, corresponding to an increase in the number of models trained. 2. Models of inferior quality can unduly impact the overall performance of the ensemble, even with the presence of highly competent models. The paper's contribution, termed MAGNAT, aligns with the concept of ensemble pruning, offering an effective and resilient consensus method for ship classification systems. Importantly, the proposed MAGNAT system allows for potential application in various maritime scenarios, including collision avoidance and route optimization.

3 MAGNAT

3.1 Principle

Figure 1 illustrates the schematic representation of MAGNAT. The maritime authority acquires and stores historical ship operation images within the ship image database. Multiple CNN models, distinguished by various configurations, are trained using the ship image database. An intelligent aggregation process is employed to selectively remove irrelevant and redundant CNN models from the ensemble. Subsequently, a drone is deployed at the port, tasked with capturing ship images from various angles. These images are then analyzed through the previously trained ensemble model. The analyzer is responsible for generating a concise situation report, alerting the authorities if abnormal maritime behaviors are identified.

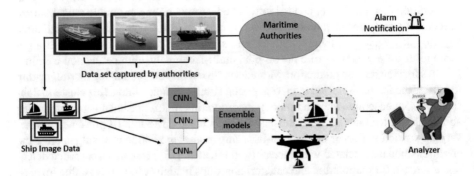

Fig. 1. MAGNAT Framework

3.2 Model Architecture Design

In this section, the author will focus on a well-known deep-learning architecture for ship classification, namely CNN. Description of the common steps for creating the CNN architectures used in the training:

Input Layer. The input is a 3-channel image with a width and height of 224 pixels: Input $\in$ R224 $\times$ 224 $\times$ 3.

Convolutional Layers. The convolutional operation is performed with k filters of size 3×3 and with a ReLU activation function. It is defined as follows:

$$C_i(X) = \text{ReLU}(W_i * X + b_i)$$

W_i is the weight matrix of the i-th fully connected layer. X is the input feature vector. b_i is bias vector of the i-th fully connected layer.

Max-Pooling Layers. Max-pooling is applied after each set of convolutional layers to downsample the feature maps. It is represented as P_i, where i is the layer number.

$$P_i(X) = \text{max}(\text{stride} \times \text{stride regions in} X)$$

Fully Connected Layers. After the convolutional and max-pooling layers, there are three fully connected layers. The fully connected operation for the i-th layer is represented as F_i:

$$F_i(X) = \text{ReLU}(W_i X + b_i)$$

The final output layer is represented as O:

$$O(X) = \text{Softmax}(W_o X + b_o)$$

W_o is the weight matrix of the output layer. X is the input feature vector. b_o is the bias vector of the output layer. Several CNN models are created by varying the number of convolution layers, the number of max-pooling layers, and the number of fully connected layers. As a result, a set of n models is designed $M = \{M_1, M_2...M_n\}$.

3.3 Intelligent Aggregation

In this section, the author will propose an intelligent aggregation of the ensemble model M already created in the previous part. In contrast to existing solutions, the proposed aggregation suggested in this paper is performed on the set of nonredundant models in the ensemble. The author defines the set of non-redundant models, noting M^* by the representative models of all models in M. The author notes $conv(M_i)$ by the set of convolution layers of the model M_i. This includes all the specifications of the convolution layer such as the number of filters, filter size, and the size of the feature map. A non-redundant set M^* is defined by the minimal subset of models that contains a maximum number of convolution layers with the same number of filters, filter size, and feature map. The target is to determine this subset M^*. The process begins by highlighting the specifications of all models. It means, the author will compute $conv(M_i)$, for all model M_i in M. A greedy search algorithm is employed to build the model search tree and eliminate models that share the same specifications as other models but have fewer convolution layers. This process will be repeated for all subsets of models in M. Following the selection of the minimal subset M^*, the voting mechanism is applied to determine the final output.

4 Performance Evaluation

To evaluate MAGNAT, intensive evaluation has been carried using the ship dataset[1]. It contains 6252 images in train and 2680 images in test data with five different classes: cargo, military, carrier, cruise, and tankers. The author also used a common metric for evaluating ship classification, "F1-score", which is a combination of precision and recall and is calculated as:

$$F1 - score = \frac{2 \times precision \times recall}{precision + recall}$$

where:

1. Precision is the number of correctly classified ships (true positives) divided by the total number of ships classified as ships (true positives + false positives).
2. Recall is the number of correctly classified ships (true positives) divided by the total number of actual ships in the dataset (true positives + false negatives).

The F1-score provides a balance between precision and recall, offering a single value that reflects the classifier's overall performance in correctly identifying ships while minimizing false positives and false negatives. The author compares MAGNAT with MFELCM, MetaBoost, and Ensemble CNN which are considered state-of-the-art ship classification models.

Figure 2 presents the F1-score value of MAGNAT and the baseline solutions MFELCM, MetaBoost, and Ensemble CNN. When varying the number of epochs from 10 to 500, the accuracy of MAGNAT is better than the other baseline methods. These results are explained by the fact that MAGNAT only uses the non-redundant models in the ensemble for inference. In other words, MAGNAT uses only the models that contribute positively to the learning process, whereas other solutions use both non-redundant and redundant models. Figure 3 presents the inference runtime of MAGNAT and the baseline solutions MFELCM, MetaBoost, and Ensemble CNN. When varying the percentage of training data from 20% to 100%, MAGNAT is faster than others. Figure 4 validates these results by showing the memory usage of MAGNAT and the baseline solutions when varying the number of models from 1 to 5. These results are reached because MAGNAT uses a subset of models in the ensemble whereas the other solutions use all models in the ensemble during the aggregation phase.

---

[1] https://www.kaggle.com/code/teeyee314/classification-of-ship-images/.

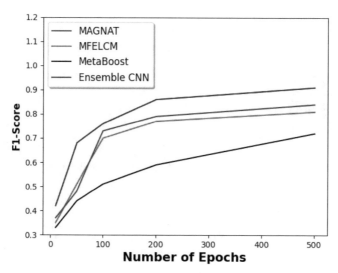

Fig. 2. Accuracy Performance

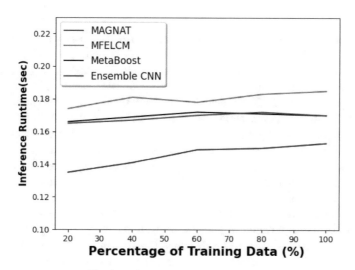

Fig. 3. Inference Runtime Performance

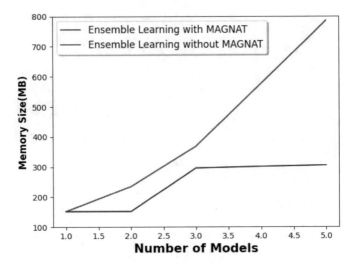

Fig. 4. Memory Usage Performance

5   Discussions

The designed MAGNAT system will bring several benefits to maritime management applications. MAGNAT will promote innovation in autonomous surveillance technology by investing in research and development to enhance the capabilities of surveillance drones and integrate them into maritime infrastructure that can contribute to sustainable development. MAGNAT can enhance maritime security, promote the rule of law at sea, and facilitate the prosecution of illegal activities. It can also help in protecting coastal communities by ensuring the safety of ships entering and leaving ports. This designed system will be ushered in a transformative era in maritime operations. It promises substantial advantages, including expanded coverage, real-time data acquisition, and a notable reduction in human risk. However, this innovative approach also introduces a distinct set of safety and security concerns that necessitate careful consideration and the development of effective mitigation strategies. Safety-wise, the risk of collisions, be it due to technical glitches, navigation errors, or unexpected obstacles, looms as a pressing issue. Adverse weather conditions can impact flight stability and sensor performance. On the security front, the interconnected nature of autonomous drones exposes them to cybersecurity vulnerabilities and potential data breaches, especially as they handle sensitive information. Unauthorized access or interference by malicious actors is another concern, as is safeguarding the physical security of these drones. Navigating these safety and security challenges is essential to harness the full potential of the designed system while ensuring the safety and confidentiality of maritime operations.

6 Conclusion

This paper introduces a novel framework for maritime management using an ensemble learning system. Several convolution neural network models have been used in the training process. Instead of exploring all models in the ensemble, this paper introduces a new aggregation strategy to prune redundant models and retain only those that contribute positively to the learning process. To validate the designed framework, intensive experiments have been carried out using a well-known ship dataset. The results reveal the superiority of the method proposed in this paper over the baseline ensemble solutions. As a potential avenue for future work, the application of other deep learning architectures, such as visual transformers [16, 17], is considered in the context of the model ensemble. Exploring other maritime computer vision tasks, such as object detection [18], and segmentation [19], is also part of the future agenda.

Acknowledgement. This work is co-funded by EU Horizon Europe under the project entitled "Smart Maritime and Underwater Guardian (SMAUG)" with grant number 101121129.

References

1. Hsu, Y.-C.: Assessment of criteria of ship classification societies. Marit. Policy Manag. **50**(7), 980–1004 (2023)
2. Wang, Y., Liu, J., Liu, R.W., Liu, Y., Yuan, Z.: Data-driven methods for detection of abnormal ship behavior: progress and trends. Ocean Eng. **271**, 113673 (2023)
3. Antaõ, P., Sun, S., Teixeira, A., Soares, C.G.: Quantitative assessment of ship collision risk influencing factors from worldwide accident and fleet data. Reliab. Eng. Syst. Saf. **234**, 109166 (2023)
4. Yasir, M., et al.: Ship detection based on deep learning using sar imagery: a systematic literature review. Soft. Comput. **27**(1), 63–84 (2023)
5. Li, J., Chen, J., Cheng, P., Yu, Z., Yu, L., Chi, C.: A survey on deep-learning-based real-time sar ship detection. IEEE J. Sel. Top. Appl. Earth Observations Remote Sens. **16**, 3218–3247 (2023)
6. Yang, Y., Lv, H., Chen, N.: A survey on ensemble learning under the era of deep learning. Artif. Intell. Rev. **56**(6), 5545–5589 (2023)
7. Campagner, A., Ciucci, D., Cabitza, F.: Aggregation models in ensemble learning: a large-scale comparison. Inf. Fusion **90**, 241–252 (2023)
8. Ganaie, M.A., Hu, M., Malik, A., Tanveer, M., Suganthan, P.: Ensemble deep learning: a review. Eng. Appl. Artif. Intell. **115**, 105151 (2022)
9. Yan, Z., Song, X., Yang, L., Wang, Y.: Ship classification in synthetic aperture radar images based on multiple classifiers ensemble learning and automatic identification system data transfer learning. Remote Sens. **14**(21), 5288 (2022)
10. Wang, Y., Yang, L., Song, X., Chen, Q., Yan, Z.: A multi-feature ensemble learning classification method for ship classification with space-based ais data. Appl. Sci. **11**(21), 10336 (2021)
11. Salem, M.H., Li, Y., Liu, Z., AbdelTawab, A.M.: A transfer learning and optimized cnn based maritime vessel classification system. Appl. Sci. **13**(3), 1912 (2023)
12. Zheng, H., Hu, Z., Liu, J., Huang, Y., Zheng, M.: Metaboost: a novel heterogeneous dcnns ensemble network with two-stage filtration for sar ship classification. IEEE Geosci. Remote Sens. Lett. **19**, 1–5 (2022)

13. Gao, J., Chi, M., Zhihui, H.: Energy consumption optimization of Inland sea ships based on operation data and ensemble learning. Math. Problems Eng. **2022**, 1–13 (2022). https://doi.org/10.1155/2022/9231782

14. Wei, Y., Chen, Z., Zhao, C., Chen, X., He, J., Zhang, C.: A time-varying ensemblemodel for ship motion prediction based on feature selection and clustering methods. Ocean Eng. **270**, 113659 (2023)

15. Liu, Y., Zhang, R., Deng, R., Zhao, J.: Ship detection and classification based oncascaded detection of hull and wake from optical satellite remote sensing imagery. GIScience Remote Sens. **60**(1), 2196159 (2023)

16. Liu, Y., et al.: A survey of visual transformers. IEEE Trans. Neural Networks Learn. Syst. 1–21 (2023)https://doi.org/10.1109/TNNLS.2022.3227717

17. Chen, S., et al.: Adaptformer: adapting vision transformers for scalable visual recognition. Adv. Neural. Inf. Process. Syst. **35**, 16664–16678 (2022)

18. Zhao, H., Zhang, H., Zhao, Y.: Yolov7-sea: object detection of maritime uav imagesbased on improved yolov7. In: Proceedings of the IEEE/CVF Winter Conference on Applications of Computer Vision, pp. 233–238 (2023)

19. Chen, X., Wu, X., Prasad, D.K., Wu, B., Postolache, O., Yang, Y.: Pixel-wise shipidentification from maritime images via a semantic segmentation model. IEEE Sens. J. **22**(18), 18180–18191 (2022)

Stock Market Prediction: Integrating Explainable AI with Conv2D Models for Candlestick Image Analysis

Joao Paulo Euko[1]([✉]), Flavio Santos[2], and Paulo Novais[3]

[1] UFS, Computer Science Department, São Cristóvão, Brazil
joaopeuko@gmail.com
[2] UFPE, IT Center, Recife, Brazil
[3] University of Minho, ISLAB, Braga, Portugal

Abstract. The term "candlestick" refers to a graphical representation of price movements in financial markets. Predicting candlesticks is crucial for anticipating market trends and making informed investment decisions. Interpretability is essential to understand the rationale behind model predictions, particularly in complex financial environments. In conclusion, this research leverages explainability techniques like SHAP and GRAD-CAM to improve the interpretability of Conv2D models for candlestick prediction in stock market images.

Keywords: xAI · deep learning · stock market

1 Introduction

In the dynamic, fast-paced financial market, where every second counts, having information and understanding why the market is moving in a certain direction can be immensely valuable. Information is the key to success, and decoding the nuances of the stock market presents a formidable challenge. The complexity of stock market time series, influenced by emotions, economic indices, politics, wars, and various other subtle factors that are difficult to discern, demands a need for understanding.

In this intricate dynamic world, where every second has a price, the significance of time series emerges. This form of data aggregates the temporal evolution of a stock market, taking into consideration price movement, emotions, indicators, and outsourced information. Being able to understand and predict these sequential patterns is the key for assertive decision-making in the landscape of the financial market. When we go further in this challenge, the journey begins by exploring deep learning applied to image time series forecasting, an incredible tool to uncover patterns that humans struggle to identify or that take more time than should be allocated.

© The Author(s), under exclusive license to Springer Nature Switzerland AG 2024
Á. Rocha et al. (Eds.): WorldCIST 2024, LNNS 986, pp. 13–23, 2024.
https://doi.org/10.1007/978-3-031-60218-4_2

With this understanding in mind, this paper aims to make time series prediction for the financial market clearer and more understandable. It explores deep learning applied to time series forecasting in an effort to predict the patterns generated by such information and explain these patterns.

Time series data is present in various aspects of our lives, and the ability to predict events based on this data allows humanity to better manage resources. For example, by monitoring temperature readings, sunlight intensity, and average rainfall, humans have been able to understand the seasons of the year and determine the optimal timing for planting and harvesting crops. The use of sophisticated computers and Artificial Intelligence has further enhanced this knowledge. However, it's important to note that such advanced technology was not available when humans initially discovered these patterns.

While it is desirable for all problems to be as straightforward and simple as predicting the seasons of the year, we understand that this particular problem has been studied for centuries. Extensive research and analysis have contributed to a better understanding of when environmental events will occur with greater precision. We have been able to filter out indicators that were initially thought to be useful in predicting seasons and have identified the most relevant ones.

With this perspective in mind, the goal is to focus on forecasting the stock market. The aim is to minimize errors when predicting if an asset is going up or down. This problem requires considering various indicators, such as price moving averages, economic indicators, political indicators, and any other indicators that traders or managers find relevant to their specific problem or asset. However, in our approach, we will solely concentrate on OLHC (Open, Low, High, Close) prices. As much as we wish the problem could be straightforward, it is not that simple. If it were, everyone would be able to predict share prices before they happen, and that wouldn't be very interesting. Throughout the centuries, investors have been trying to forecast the next price movement in the stock market by utilizing the indicators mentioned earlier. Various strategies have been developed to aid in price prediction.

Due to the rapid and complex nature of price movements, we believe it is essential to have a tool that is capable of processing information faster than the human eye and can capture patterns that may be difficult for humans to discern or require extensive training to identify. This is where deep learning comes into play. In the upcoming sections, we will discuss how deep learning, an artificial intelligence technique, has been proven to effectively predict price time series. However, it's important to note that the discussion will focus on historical data, and the real-time aspect will not be considered at this moment.

Finally, the goal is to develop a method that can explain the decisions made during the forecasting process, exclusively utilizing OLHC data. The idea is to identify and focus only on the most crucial indicators that accurately describe price formation. By understanding what is relevant for predicting a specific problem, performance can be significantly improved.

1.1   Objectives of the Research

Explainable Artificial Intelligence (xAI) represents a sophisticated subfield devoted to enhancing model interpretability. Specifically applied to stock market patterns in our context, xAI aims to determine the significance of features in the prediction process. This subfield involves assessing the relevance of each input feature, facilitating manual or automated filtering or validation based on a hyperparameter threshold. Notably, approaches such as GRAD-CAM and SHAP show promise in meeting the demand for comprehensibility among both humans and machines. By offering insights into the importance of features, they enable us to strategically filter and focus on the most pertinent information for improved interpretability.

Our proposal involves investigating whether xAI can effectively enhance the interpretability of Conv2D models, improve prediction performance, and clarify which features are being more considered. As of now, there are already notable works in the field that offer explainability, such as SHAP and LIME, and we aim to conduct further research to explore the best alternatives, considering both speed and prediction performance.

We are exploring the integration of Conv2D models and explainability techniques, specifically SHAP and GRAD-CAM, to enhance the interpretability of these models in predicting candlestick patterns. Our primary focus is on delivering more answers than questions.

2   Related Work

Explainable Artificial Intelligence (xAI). xAI is a complex field that integrates various methods to simplify the decision-making processes of black box models. According to [1], "AI applications must not only perform well in terms of classification metrics but also need to be trustworthy and transparent." xAI plays a crucial role in identifying significant features in model predictions, enhancing clarity in interpreting results.

Breaking down the complexities of these models, xAI gives us useful insights into what affects predictions. This makes it simpler for people to understand how decisions are made. Another good thing about xAI is that it makes these complex models more user-friendly, helping us humans understand them better.

Investigating Explainability Methods in Recurrent Neural Network Architectures for Financial Time Series Data. My initial exploration focused on delving into existing research on xAI within the realm of time series forecasting. We found the paper by [4] to be a crucial starting point for our investigation.

The insights derived from [4] were particularly encouraging. The authors concluded, "The results show that these methods are transferable to the financial forecasting sector, but further confirmation is needed for a more sophisticated hybrid prediction system" [4]. This work not only validated the applicability of Explainability methods in time series forecasting but also emphasized the potential in the context of the stock market. Consequently, it served as a solid foundation, supporting the assertion that employing xAI in financial time series prediction is a plausible and promising endeavor.

Explainable AI and Adoption of Financial Algorithmic Advisors: An Experimental Study. Motivated by a prior study that highlighted the potential of Explainable Artificial Intelligence (xAI) in financial time series forecasting, an exploration ensued into the degree of trust individuals place in explanations provided by artificial intelligence. A relevant study by [2] was identified, involving a game where participants made decisions regarding the quantity of lemons to purchase for lemonade production.

As the game progressed, participants received various reports, including one generated by an artificial intelligence system. The findings from [2] unveiled a significant trend - individuals displayed a willingness to trust an artificial intelligence report when it could furnish clear and understandable explanations for the decisions made. This observation not only enriches our understanding of human-AI interaction but also establishes a crucial context for the ongoing investigation into trust in AI-generated explanations.

The Best Way to Select Features? Comparing MDA, LIME and SHAP. Now that we are aware of the availability of several free and open-source xAI frameworks in the market, we need to decide which ones to use. After reading the [6] paper, we decided to use SHAP and GRAD-CAM for several reasons. According to the [6], SHAP demonstrated stability and consistently provided reliable results. Additionally, the SHAP framework is compatible with various deep learning frameworks, as is GRAD-CAM, such as TensorFlow. Although we considered several other options like LIME, in the end, the advantages of SHAP and GRAD-CAM outweighed the alternatives.

SHAP. Since the previous subsections, we mentioned that we will be using SHAP. What precisely constitutes SHAP?

The SHAP framework operates by taking a trained model and iterating through all its features. It evaluates the importance of each feature by considering their individual contributions and averaging the differences between the actual and expected values. SHAP incorporates properties such as "local accuracy, missingness, and consistency" [5]. Moreover, it achieves this by constraining attention.

This research explores different variants of SHAP and compares them with LIME and linear layer models to explain feature importance. Considering the availability of information and its suitability for explainable AI, SHAP has been chosen as the framework for this work.

3   Method

The analysis method involves collecting daily OHLC (Open, High, Low, Close) data from the S&P 500, transforming it into candlestick images, and grouping the data into blocks of 10 candles. These grouped images are labeled based on the closing price of the 10th candle relative to the 11th candle. The primary focus is on Conv2D models for predicting price movements. Explainability is enhanced using SHAP and GRAD-CAM techniques. The main evaluation metric is accuracy, comparing predicted labels with ground truth. The experimental design includes training Conv2D models and applying explainability techniques.

4   Experiments and Results

This section presents the methodology, dataset collection, model architecture, explainability techniques (SHAP and GRAD-CAM), evaluation metrics (accuracy), and the experimental design.

4.1   Dataset

For the image challenge, data collection involves gathering the past years of daily OHLC (Open, High, Low, Close) data from the S&P 500 on NASDAQ, the USA stock market index representing the most traded companies on NASDAQ. Subsequently, the collected data undergoes preprocessing, where it is transformed into candlesticks-a visual representation of price movements on a chart. This preprocessing is accomplished by converting the CSV containing historical prices into a candlestick chart, cropping the image, and saving it for later use. Following the methodology outlined in [3], the data is grouped into blocks of 10 candles. This grouping allows for the observation of price movements, and each group is labeled based on whether the closing price of the sixth candle is above or below the closing price of the tenth candle.

Fig. 1. Visualization of Financial Data: A 10-candle image representing grouped price movements in the S&P 500 on NASDAQ, where each block of candles captures the market dynamics over a specific time period.

In Fig. 1, we present an example of the image we obtained. Each bar, referred to as a candlestick, conveys information about the open, close, high, and low prices. To enhance comprehension, the candlesticks are color-coded: red indicates that the price closed below its open, while green indicates closure above the open. The thin line atop the larger body indicates the maximum price at that moment, and the thin line at the bottom denotes the minimum price at that moment.

Our dataset spans from August 30, 2013, to August 29, 2023, with each candle representing the price movement for a single day.

In Fig. 1, we display 10 d.

4.2   Model Architecture

For the image challenge, the primary focus will be on Conv2D models. These models will undergo training and comparative analysis to determine their performance in predicting price movements of stock market candlestick images.

The architectural details of the Conv2D models that will be utilized in the experiments are presented below (Fig. 2).

Fig. 2. Model Architecture Diagram

4.3   Explainable AI Techniques

In this section, we will focus on employing SHAP and GRAD-CAM as explainability techniques for Conv2D models. Some challenges are anticipated for these techniques, and one notable challenge is the interoperability of the frameworks. While some of the chosen frameworks work seamlessly together, others may have conflicts, posing potential challenges in the future.

The methodology for both Explainable Artificial Intelligence (xAI) techniques is similar. A Conv2D model will be trained, and subsequently, SHAP and GRAD-CAM will explain the image, elucidating why it believes the market will move up or down. Armed with this information, features will be selected, such as determining the optimal background size for the chart. This process aims to enhance the interpretability of the Conv2D model's predictions. With unnecessary information removed, the model will be retrained using the same architecture as mentioned earlier. The performance will be compared to assess whether a model, enriched with insights from SHAP and GRAD-CAM, demonstrates improved interpretability and prediction performance. Moreover, it will reveal which features are considered significant for the prediction, providing insights into future decision-making processes.

4.4   Evaluation Metrics

The most important metric and the principal one will be accuracy, as the objective is to predict the next candlestick closing price. The accuracy will be calculated by comparing the predicted label with ground truth. The ground truth is the label of the next candlestick closing price. The accuracy will be calculated by dividing the number of correct predictions by the total number of predictions.

4.5   Experimental Design

The experiment consists of gathering historical data, converting it to images containing five candles each, and labeling it with the next candlestick closing price. After that, the data will be split into train and validation sets. The train set will be used to train the Conv2D model, and the test set will be used to evaluate the model. The accuracy will be calculated by comparing the predicted label with ground truth, representing the label of the next candlestick closing price.

Afterward, SHAP and GRAD-CAM will be applied as explainability techniques to the Conv2D model, and the model will be retrained. The accuracy will be calculated again and compared with the accuracy of the original Conv2D model.

Both images from xAI from non-filtered and filtered, as well as the accuracy level, will be compared.

4.6 Model Training and Accuracy

The Conv2D models were trained using images generated from S&P500 OHLC data. To facilitate this, the tabular data underwent a preprocessing step where it was transformed into images, as illustrated in Fig. 3. Each image in the dataset represents a group of price movements within a block of 10 candles.

The training objective was to predict whether the closing price of the 11th candle would be higher or lower than the closing price of the 10th candle. The models were evaluated based on their accuracy in predicting this directional movement.

Fig. 3. 10-candle image for price movement prediction in the S&P 500 on NASDAQ.

The models achieved an accuracy close to 100% (see Fig. 4) and had an error close to zero (see Fig. 4) during training. However, in the validation data, the accuracy remained around 55%.

Fig. 4. Accuracy (a) and Loss (b) charts during training, with the x-axis representing the number of epochs and the y-axis indicating the corresponding metric.

5 Discussion

5.1 Explainability

On the previous section, we observed the Fig. 3, utilized for predicting the next candle, yielding a prediction accuracy of 99.09% for a green candle. Remarkably, the prediction proved accurate, as the subsequent candle indeed turned out to be green.

Upon closer inspection, the price trend displayed a succession of red candles, suggesting a potential market decline. To delve deeper into the market dynamics during this period, Fig. 5 provides insights.

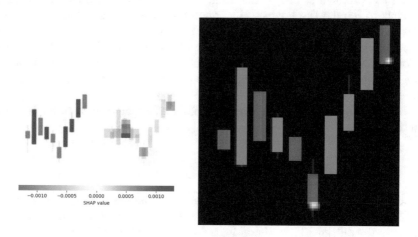

Fig. 5. Explainability. Interpretation of Prediction in Each Image.

It is possible to see in Fig. 5 three images. The leftmost one is the original image, sharing information with the SHAP visualization, and on the rightmost is the GradCam visualization. It is interesting to note that in both methods, the most important region is almost the same. The green candle in the middle of the image, representing the price reversion for that moment, had its open price checked by the models, and the open price on the last candle was also important for the correct price prediction. SHAP also had a good amount of importance on the green candle before the one in the middle.

5.2 Insights Gained

The results obtained from the experiments revealed that the Conv2D models were able to predict the next candlestick closing direction with high accuracy in training. However, the accuracy of the models in predicting the direction of the next candlestick closing price was significantly lower in validation.

It is understandable since it is a very complex task to predict the next candlestick closing price direction. The models were able to learn the patterns in the training data, but they were not able to generalize well to the validation data.

However, the ability to explain the market was found to be satisfactory and to indicate that the model is looking at the right places to make the prediction.

5.3 Limitations and Future Work

It is important to take into consideration some of the limitations. In the early stages of the research, it was possible to train the model with larger images (1025, 1025), which led to the belief that the model would be able to predict the next candlestick closing price direction with high accuracy in validation. However, when the problem scaled in the number of images and involved more complex models to handle the images, the computer used was not able to support the size of the data, as well as the available clouds.

The model might perform poorly in validation, but the explainability intuition for the model was impressive. Future work would replicate the same experiment with a larger computer and more data to see if the model would be able to generalize better.

Also, it would be interesting to explore other models in order to look for better results.

6   Conclusion

In this research, we delved into the domain of the stock market, exploring the utilization of Conv2D models for predicting the next candlestick closing price direction. Additionally, we investigated the application of Explainable Artificial Intelligence (xAI) techniques to provide insights into the predictions made by these Conv2D models.

Our primary goal was to elucidate the predictions of the Conv2D models and offer a deeper understanding of why specific predictions were generated.

The results obtained from our experiments indicated that the Conv2D models exhibited high accuracy in predicting the next candlestick closing direction during training. However, their accuracy significantly decreased when predicting the direction of the next candlestick closing price in validation.

This decline in accuracy is comprehensible, given the intricate nature of predicting the next candlestick closing price direction. It also emphasizes the necessity for further investigation into potential overfitting and the exploration of more robust models.

The use of xAI produced significant results, providing satisfactory outcomes that indicated the model's ability to focus on relevant information for making predictions.

Despite the potential decrease in model performance during validation, the impressive explainability intuition underscores the significance of acknowledging the limitations faced during experimental development, particularly concerning computational power.

In conclusion, our research showcases the viability of employing xAI to explicate Conv2D models for image classification in stock market prediction. While the results warrant further investigation, they establish a solid foundational point.

Future efforts will involve replicating the experiment with an improved computational infrastructure and an expanded dataset to assess whether the model can generalize more effectively. This endeavor aims to address the encountered limitations in this research.

References

1. Angelov, P.P., Soares, E.A., Jiang, R., Arnold, N.I., Atkinson, P.M.: Explainable artificial intelligence: an analytical review. WIREs Data Min. Knowl. Discov. **11**(5), e1424 (2021). https://doi.org/10.1002/widm.1424, https://wires.onlinelibrary.wiley.com/doi/10.1002/widm.1424
2. David, D.B., Resheff, Y.S., Tron, T.: Explainable AI and Adoption of Financial Algorithmic Advisors: an Experimental Study (2021). arXiv:2101.02555 [cs]
3. Franken, G.: Applications of Content-Based Image Retrieval in Financial Trading
4. Freeborough, W., van Zyl, T.: Investigating explainability methods in recurrent neural network architectures for financial time series data. Appl. Sci. **12**(3), 1427 (2022). https://doi.org/10.3390/app12031427, https://www.mdpi.com/2076-3417/12/3/1427
5. Lundberg, S., Lee, S.I.: A Unified Approach to Interpreting Model Predictions (2017). http://arxiv.org/abs/1705.07874, arXiv:1705.07874 [cs, stat]
6. Man, X., Chan, E.P.: The best way to select features? Comparing MDA, LIME, and SHAP. J. Financ. Data Sci. **3**(1), 127–139 (2021). https://doi.org/10.3905/jfds.2020.1.047, http://jfds.pm-research.com/lookup/doi/10.3905/jfds.2020.1.047

Resources Optimization and Value-Based Prioritization for at Risk Cultural Heritage Assets Management

Ulysse Rosselet[✉] and Cédric Gaspoz

University of Applied Sciences Western Switzerland (HES-SO), HEG Arc, Neuchâtel, Switzerland
ulysse.rosselet@he-arc.ch

Abstract. The paper examines efficient post-disaster management of cultural artifacts, emphasizing the need to prioritize object treatment for resource efficiency. It focuses on multi-objective optimization considering heritage value, processing capacities, and resource availability. Highlighting gaps in disaster recovery guidelines and ICT utilization, it proposes using optimization techniques to adapt object routing based on heritage value amidst uncertainties from incomplete data. It suggests queueing theory, job shop, and batch-processing optimizations for object recovery processes. The research underscores the significance of optimizing the stabilization phase during recovery, presenting a model for prioritizing objects based on heritage value and damage. It seeks to allocate rescue budgets to objects, optimizing their stabilization within budget constraints to maximize salvage impact.

Keywords: cultural heritage recovery · object prioritization · multi-objective optimization

1 Introduction

After a disaster affecting cultural and heritage property, efficient resource allocation and prioritization are crucial for object stabilization and evacuation. Institutions face the challenge of handling diverse artifacts with varying heritage values and limited processing capacities. Effective salvage operations depend on managing and adapting priorities during recovery and stabilization phases (Kjølsen Jernæs, 2021). This paper focuses on the multi-objective optimization of the object processing flow based on their heritage value, the capacity of processing lines and the availability of resources. After presenting the specific issues of cultural assets disaster management and relevant optimization techniques in the literature review section, we will present our rescue and recovery process and model the resulting optimization problem. The discussion section will present the way these techniques can be applied and our existing efforts in validating our approach. The conclusion will open avenues for future research and synthesize our findings. The main contribution of this paper is the modeling of the resource allocation and object prioritization problem for cultural assets disaster management.

Á. Rocha et al. (Eds.): WorldCIST 2024, LNNS 986, pp. 24–33, 2024.
https://doi.org/10.1007/978-3-031-60218-4_3

2 Literature Review

Cultural and Heritage Assets Disaster Management. In crisis management, distinguishing between disaster, hazard, and risk is crucial (Gravley, 2001; Mayner & Arbon, 2015). Hazards are potential threats, disasters are events affecting people or property, and risk combines event likelihood and outcomes (ISO; Drennan, McConnell, & Stark, 2014).

Disaster management involves pre-crisis, crisis, and post-crisis phases (Lettieri, Masella, & Radaelli, 2009), with stages like mitigation, preparedness, response, and recovery (Lettieri et al., 2009; Mohan & Mittal, 2020; Sakurai & Murayama, 2019). Information systems play a key role in these stages.

Heritage collection management, including preservation and inventory, has standardized methods (Dudley & Wilkinson, 1958; Arvanitis & Tythacott, 2014 Matassa, 2011; International Council of Museums, 2017). Protecting these collections during disasters requires ethical considerations and international conventions, yet lacks specific guidelines for individual object rescue (Coté & Carr, 2022).

Recovery involves data input, object identification, and analytics, often relying on manual systems but with potential for ICT use (Caire, 2003; Chroust, 2012; Graves, 2004). Prioritizing and adapting during recovery is crucial for salvaging heritage collections (Kjølsen Jernæs, 2021), requiring a balance in resource allocation and expert decision-making (Shan & Yan, 2017).

Optimization Techniques. Optimization aims to discover the most optimal solution within defined criteria and constraints, involving maximizing or minimizing a function while considering a set of limitations (Türkyılmaz, Şenvar, Ünal, & Bulkan, 2020). In our scenario, we strive to simultaneously optimize multiple objectives: preserving total heritage value and reducing processing time for priority objects. While achieving a global optimum is the ultimate aim, in practical scenarios, improving upon existing solutions within time or financial constraints is often deemed successful (Stork, Eiben, & Bartz-Beielstein, 2022; Caunhye et al., 2012). Optimization problems vary based on the problem type, number of objectives pursued, and data accuracy, categorized into deterministic or stochastic optimization. Local optimization algorithms, mainly gradient-based, suit problems with numerous variables but might not align well with discrete optimization tasks. Conversely, global optimization algorithms have better potential for finding the global optimum, especially with discrete features, although they come with high computational costs and limited constraint handling abilities (Venter, 2010). Given the uncertainty inherent in our context, stochastic optimization, accounting for unpredictability in object extraction order and rate, becomes crucial. However, selecting a specific optimization method remains challenging as research indicates that context-specific considerations and hybridization of competing algorithms often yield superior results compared to individual algorithm application (Fouskakis & Draper, 2002).

In terms of optimization, sequencing and scheduling the treatment operations in the case of post disaster rescue and recovery of heritage objects can be formulated as different alternative problems.

First, we can formulate the problem as a budget optimization problem where each heritage object possesses a finite budget allocated for stabilization (Lee et al., 2010), while various stabilization measures come with specific costs. This optimization problem requires a strategic decision-making framework to maximize the effectiveness of stabilization efforts within these budget constraints (Soma et al., 2014). The challenge lies in selecting from a range of available stabilization techniques, each with its associated cost, while adhering to the financial limits imposed on each object. The objective is to optimize the selection of stabilization measures across all objects to achieve the most impactful stabilization outcome while staying within the prescribed budgetary constraints for each object. This formulation seeks to identify an optimal combination of stabilization measures for each object that maximizes the overall effectiveness of the stabilization process while respecting the budgetary limitations unique to each heritage item.

Second, we could consider our problem as a special case of the job shop optimization problem, that aims to find the optimum schedule for allocating shared resources over time to competing activities in order to reduce the overall time needed to complete all activities (Yamada & Nakano, 1997). During the last three decades, numerous research efforts have focused on solving this problem efficiently (Chaudhry & Khan, 2016; Türkyılmaz et al., 2020; Xie, Gao, Peng, Li, & Li, 2019). In our context, we seek to optimize at least two objectives simultaneously: maximizing the preservation of total heritage value on the one hand and minimizing the processing time for priority objects on the other. Therefore, we need to find a multi-objective optimization technique suitable for solving our variant of the flexible job shop scheduling problem.

To summarize, there's a notable lack of standardized guidelines for individual object rescue and recovery. Effective resource allocation and object routing are crucial for preserving significant objects. In the field of ICT applied to post disaster collection rescue and recovery, emergency response decision support systems (ERDSS) could be used to provide institutions with reliable and efficient tools helping them manage the rescue, recovery and stabilization of impacted collections (Pettet et al., 2022). Examples of ERDSS in the field of heritage protection comprise risk monitoring platforms such as SOS-Heritage (https://www.sos-heritage.eu/), rescue and recovery systems like Arcultura (https://arcultura.org/), and general methodologies such as Proculther (https://www.proculther.eu/). Optimizing the routing and processing of objects based on heritage value could enhance response efficiency. Yet, further research is necessary to identify and integrate optimal optimization techniques for this purpose.

3   Optimization of the Rescue and Recovery Process for Cultural Heritage Assets

In this section, we present the processes on which we will use optimization techniques, depicted in Fig. 1 and further detailed in a white paper by Rosselet et al. (2022). At the most abstract level, it can be viewed as three main phases: extraction and assessment, stabilization, reintegration.

Extraction and assessment denote the phase following the impact when affected objects are removed from the disaster site and transported to the assessment station. During this stage, the impacted assets undergo evaluation to determine the extent of damage. The stabilization phase consists of all measures required for the heritage asset to reach a stable conservation state, that is, the item is in a condition that will not further degrade, and it is ready for temporary storage and ulterior restoration or reintegration.

The reintegration phase is the moment stabilized assets can return to traditional or temporary storage locations, their data can be reconciled with the usual collection management systems and further restoration measures can be used as part of the day-to-day processes of the institution.

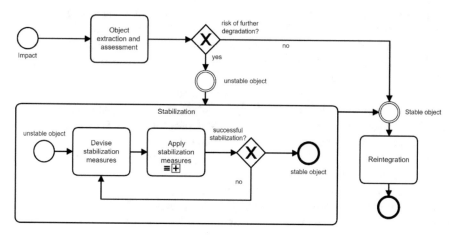

Fig. 1. Simplified object rescue and recovery process

Among these phases, our optimization effort will target specifically the stabilization phase. During this phase, the actors in charge of the rescue and recovery need to choose how to apply the different stabilization measures and how to balance the available resources. This phase is critical to avoid further degradation of the assets.

Optimization of the Stabilization Process. The first step is to model the prioritization problem and define the objective function, decision variables and set of constraints that restrict the values of the decision variables for the optimization problem.

Our goal is to maximize the value of the salvage effort under the constraint of the available processing resources. To make the best use of available resources and ensure efficient rescue, two sets of factors must be considered at object level: their priority, and their salvageability.

The objects are characterized on a 5 level priority scale comprising low, medium, high, extreme and catastrophic priority (Pedersoli & Michalski, 2016). The priority level of an object may have been determined in advance - this is often the case for high-value objects on the institution's priority lists - or it may be established when the objects are sorted during the salvaging process.

The salvageability factor depends on the level of damage sustained by the object and the extent of stabilization required. The level of damage is described by a 5-level scale: no damage, light/minor, moderate, severe, destroyed (PROCULTHER Project, 2021). The extent of stabilization required is quantified by the number of required treatments and their duration.

Optimization Problem as a Budget Allocation Problem. The optimization problem can be modeled as a budget allocation problem. Each object is assigned a rescue budget.

This rescue budget is first based on the existing object's valuation and priority. Object valuation and priority can exist beforehand (which is typically the case for insured and high priority objects) or can be established during the evaluation phase of the rescue and salvaging process.

The rescue budget determines the stabilization measures the object will undergo and its priority for treatment. Each stabilization measure has a marginally increasing cost, to reflect the fact that a stabilization measure's availability decreases as it reaches its saturation level. During the stabilization phase, there exist different alternatives for stabilization measures that vary in both cost and effectiveness in preserving an object's authenticity and integrity.

The resource allocation process can be considered based on the total rescue cost for an object in comparison to its rescue budget. The objective is to maximize the value of the salvage effort under the constraint of the available processing resources.

The optimization problem is modeled as follows:

- B_i as the budget allocated for object i
- M_{ij} representing stabilization measures for object i ($j = 1, 2, \ldots, m_i$ for the ith object)
- P_{ij} denoting the priority of each stabilization measures M_{ij} for object i
- C_{ij} as the cost associated with implementing M_{ij} for object i

Decision Variables

- Let x_{ij} be a binary variable indicating whether a specific stabilization measure M_{ij} is selected for object i.
- $x_{ij} = \begin{cases} 1 & \text{if stabilization measure } M_{ij} \text{ is selected for object } i \\ 0 & \text{otherwise} \end{cases}$

Objective Function

Maximize the overall impact or priority of selected stabilization measures for each object:

- $\max \sum_{i=1}^{N} \sum_{j=1}^{m_i} P_{ij} \cdot x_{ij}$

Constraints

Budget constraint for each object: the total cost of selected measures for each object should not exceed its allocated budget:

- $\sum_{j=1}^{m_i} C_{ij} \cdot x_{ij} \leq B_i$ for $i = 1, 2, \ldots, N$

Binary variable constraint: x_{ij} should be binary (either 0 or 1) for all i and j:

- $x_{ij} \in 0, 1$ for $i = 1, 2, \ldots, N$ and $j = 1, 2, \ldots, m_i$

These constraints ensure that for each object, the selected stabilization measures do not surpass the allocated budget, while each x_{ij} remains a binary decision variable.

This formulation allows for optimization tailored to each individual object, maximizing the overall priority or impact of stabilization measures within their respective budgets. Adjustments can be made based on specific costs, priorities, and constraints associated with the stabilization measures for each object.

Optimization as a Special Case of the Job Shop Optimization Problem. The second modelling we investigated it considering our problem as a special case of the job shop optimization problem, that aims to find the optimum schedule for allocating shared resources over time to competing activities in order to reduce the overall time needed to complete all activities (Yamada & Nakano, 1997; Özgüven et al., 2010).

In this type of problem, each heritage object would represent a job that requires undergoing various stabilization measures for preservation. The different alternative stabilization measures or tasks required for the preservation of heritage objects are represented as machines. To reflect the variety of available methods for object stabilization, we define multiple alternative stabilization measures for each task, considering variations in cost and effectiveness in preserving authenticity and integrity. These alternatives represent different options for achieving the same stabilization goal.
The optimization problem is formulated as follows:

- i objects
- j stabilization measures
- J the set of objects requiring stabilization
- B_i as the budget allocated for object i
- V_i as the preservation value of heritage object i
- M_i representing the ordered set of stabilization measures for object i where $M_{if_{(i)}}$ is the first and $M_{il_{(i)}}$ is the last element.
- P_j as the set of processing lines on which operation j can be processed
- t_{ijk} as the time required to implement M_{ij} for object i on processing line k
- C_{ijk} as the cost associated with implementing M_{ij} on processing line k

Decision variables:

- $x_{ijk} = \begin{cases} 1 \text{ if processing line } k \text{ is selected for stabilization measure } M_{ij} \text{ for object } i \\ 0 \qquad\qquad\qquad\qquad\qquad \text{otherwise} \end{cases}$
- $s_{ijk} = $ the starting time of operation O_{ij} on processing line k
- $c_{ijk} = $ the completion time of operation O_{ij} on processing line k
- $c_i = $ the completion time for stabilization of object i

The objective functions are:

Maximize Preservation Value: Maximize the total value preserved across all heritage objects considering their historical, cultural, and monetary significance.

- $\max \sum_i V_i$

Minimize Processing Time: Minimize the time taken to complete priority objects' stabilization measures, ensuring timely preservation.

- $\min \sum_{i=1}^{N} c_i$

Minimize Cost: Minimize the overall cost incurred in selecting stabilization measures while staying within allocated budgets.

- $\min \sum_{i=1}^{N} \sum_{j=1}^{m_i} C_{ij} \cdot x_{ij}$

Constraints:

Budget: Each heritage object is allocated a rescue budget that should not be exceeded during the stabilization process.

- $\sum_{k \in P_j} C_{ijk} \cdot x_{ijk} \leq B_i \, \forall i \in J, \, \forall j \in M_i$

Precedence: Certain stabilization measures must be completed before others can start.

- $\sum_{k \in P_j} s_{ijk} \geq \sum_{k \in P_j} c_{ij-1,k} \, \forall i \in J, \, \forall j \in M_i - \{M_{if_{(i)}}\}$

Time: Specific time frames within which priority objects need to undergo stabilization.

- $c_i \geq \sum_{k \in P_j} C_{i,M_{il_{(i)}},k} \, \forall i \in j$

This modelling better captures the stakes of the rescue and recovery process. Solving this refined job shop scheduling problem considering alternative stabilization measures involves finding an optimal schedule that maximizes preservation value, minimizes processing time for priority objects, and minimizes costs by selecting the most suitable and cost-effective alternatives for stabilization measures within allocated budgets and considering the effectiveness in preserving authenticity and integrity.

4 Results and Discussion

In this paper, we have modeled the problem as a budget allocation challenge, where each heritage object's rescue budget determines the stabilization measures within financial constraints. The goal was to maximize the overall impact or priority of selected stabilization measures for each object within the allocated budget. This optimization problem could be solved by using a binary integer programming solver.

The second modelling presented the problem as a special case of the job shop optimization problem, aiming to find an optimum schedule for allocating shared resources over time to competing activities. Each heritage object represented a job, and alternative stabilization measures were treated as machines. The goals are to maximize preservation value, reduce processing time, and minimize costs. To solve this, mixed integer linear programming could be used (Özgüven et al., 2010).

The current study presents several pertinent limitations. Firstly, the optimization process lacks consideration of uncertainties, which are pivotal in influencing its efficacy. Our problem may call for the use of stochastic techniques because it involves a significant amount of uncertainty (Powell, 2019), mainly because data required to solve the optimization problem is not known completely in advance: the complete inventory of impacted objects may not exist or may have been destroyed, new objects are continuously discovered and extracted from the damaged areas, processing times can vary between objects, required stabilization treatments can be modified to tailor the stabilization process to the evolution of an object's state. Secondly, while optimizing the stabilization process might yield cost savings, there remains a crucial need to balance these against the overall expenses incurred in global salvage operations. Thirdly, the realization of the optimization's full potential efficacy remains uncertain due to potential impediments such as bottlenecks or organizational hurdles within the salvage efforts, which might impede the anticipated benefits.

5   Conclusion

The paper explores the effective management of cultural and heritage collections after a disaster, emphasizing prioritizing object treatment for resource efficiency and optimal care. Institutions face diverse artifacts with varying heritage value and limited processing capacities, necessitating efficient prioritization strategies. The study focuses on multi-objective optimization of object processing considering heritage value, processing line capacity, and resource availability. It delves into disaster management phases, the lack of standardized guidelines for individual object recovery, and the role of information technology in aiding mitigation, preparedness, response, and recovery. Moreover, it discusses the significance of object traceability for preservation and legal aspects.

Key gaps identified include the absence of operational guidelines for post-disaster recovery at an individual object level and insufficient recommendations for implementing ICT in this context. The paper proposes utilizing optimization techniques to dynamically adapt object routing and processing based on heritage value, addressing uncertainties arising from incomplete data. Techniques like queueing theory, job shop optimization, and batch-processing optimization are considered for optimizing object recovery processes. The research section advocates for stochastic optimization techniques due to their adaptability and faster solutions.

The specific focus of the paper lies in optimizing the stabilization phase during object recovery. It presents a model for prioritizing objects based on their heritage value and the extent of damage suffered. The optimization problem revolves around allocating rescue budgets to objects, determining their stabilization measures within the budget constraints, and maximizing the salvage effort's overall impact. The formulated decision variables and constraints aim to optimize each object's treatment while managing resource limitations effectively. This approach ensures tailored optimization for individual objects, maximizing the impact of stabilization measures within allocated budgets.

While the job shop scheduling variant more accurately reflects the constraints involved in the rescue and recovery of real-world heritage objects, the potential benefits must be weighed against the increased complexity of this approach. Meanwhile, the budget allocation optimization technique may yield satisfactory results in practical applications, where other factors outside the scope of this paper may have a greater influence on the rescue and recovery outcomes. In further research we would favor the use of heuristic or metaheuristic techniques due to their higher applicability and their ability to present fast sufficient solutions (Stork et al., 2022). Among these techniques, we would investigate optimization algorithms inspired by natural, human, social, or physical phenomena (Zhang et al., 2018).

Acknowledgement. We are grateful to the Hasler Foundation which provided partial funding for this work under grant number 22070.

References

Arvanitis, D.K., Tythacott, D.L.: Museums and Restitution: New Practices. Ashgate Publishing Ltd, New Approaches (2014)

Caunhye, A.M., Nie, X., Pokharel, S.: Optimization models in emergency logistics: a literature review. Socioecon. Plann. Sci. **46**, 4–13 (2012)

Chaudhry, I.A., Khan, A.A.: A research survey: review of flexible job shop scheduling techniques. Int. Trans. Oper. Res. **23**, 551–591 (2016)

Coté, N., Carr, J.: Ready or not: emergency planning in Midwestern museums. Mus. Manage. Curatorship **37**, 116–136 (2022)

Drennan, L.T., McConnell, A., Stark, A.: Risk and Crisis Management in the Public Sector. Routledge (2014)

Dudley, D. H., & Wilkinson, I. B. (1958). *Museum Registration Methods*. American Association of Museums

Fouskakis, D., Draper, D.: Stochastic optimization: a review. Int. Stat. Rev. **70**, 315–349 (2002)

Graves, R.: Key technologies for emergency response. In: First International Workshop on Information Systems for Crisis Response and Management ISCRAM2004, Brussels (2004)

Gravley, D.: Risk, Hazard and Disaster. University of Canterbury, New Zealand (2001)

International Council of Museums: ICOM Code of Ethics for Museums. ICOM, Paris (2017)

Lee, L.H., Chen, C., Chew, E.P., Li, J., Pujowidianto, N.A., Zhang, S.: A review of optimal computing budget allocation algorithms for simulation optimization problem. Int. J. Oper. Res. **7**(2), 19–31 (2010)

Lettieri, E., Masella, C., Radaelli, G.: Disaster management: findings from a systematic review. Disaster Prev. Manage. Int. J. **18**, 117–136 (2009)

Matassa, F.: Museum Collections Management. Facet (2011). https://doi.org/10.29085/978185 6048699

Mayner, L., Arbon, P.: Defining disaster: the need for harmonisation of terminology. Australas. J. Disaster Trauma Stud. **19**, 21–26 (2015)

Mohan, P., Mittal, H.: Review of ICT usage in disaster management. Int. J. Inf. Technol. **12**, 955–962 (2020)

Özgüven, C., Özbakır, L., Yavuz, Y.: Mathematical models for job-shop scheduling problems with routing and process plan flexibility. Appl. Math. Model. **34**(6), 1539–1548 (2010)

Pedersoli, J.L.J., Michalski, S.: A guide to risk management of cultural heritage. In: International Centre for the Study of the Preservation and Restoration of Cultural Property ICCROM (2016)

Pettet, G., et al.: Designing decision support systems for emergency response: challenges and opportunities. In: 2022 Workshop on Cyber Physical Systems for Emergency Response (CPS-ER), pp. 30–35 (2022)

Powell, W.B.: A unified framework for stochastic optimization. Eur. J. Oper. Res. **275**, 795–821 (2019)

PROCULTHER Project: Key Elements of a European Methodology to Address the Protection of Cultural Heritage during Emergencies. LuoghInteriori (2021)

Rosselet, U., Jacot, T., Gaspoz, C., Gelbert-Miermon, A.: Cahier des charges pour un Dispositif Intégré d'Enregistrement et de Suivi des Objets Sinistrés [White paper]. University of applied sciences Western Switzerland (2022)

Sakurai, M., Murayama, Y.: Information technologies and disaster management – Benefits and issues. Progress Disaster Sci. **2**, 100012 (2019)

Shan, S., Yan, Q.: Emergency Response Decision Support System. Springer (2017). https://doi.org/10.1007/978-981-10-3542-5

Soma, T., Kakimura, N., Inaba, K., Kawarabayashi, K.: Optimal budget allocation: Theoretical guarantee and efficient algorithm. In: International Conference on Machine Learning, pp. 351–359 (2014). https://proceedings.mlr.press/v32/soma14.html

Stork, J., Eiben, A.E., Bartz-Beielstein, T.: A new taxonomy of global optimization algorithms. Nat. Comput. **21**, 219–242 (2022)

Türkyılmaz, A., Şenvar, Ö., Ünal, İ, Bulkan, S.: A research survey: heuristic approaches for solving multi objective flexible job shop problems. J. Intell. Manuf. **31**, 1949–1983 (2020)

Venter, G.: Review of optimization techniques (2010). https://scholar.sun.ac.za:443/handle/10019.1/14646

Xie, J., Gao, L., Peng, K., Li, X., Li, H.: Review on flexible job shop scheduling. IET Collaborative Intell. Manuf. **1**, 67–77 (2019)

Yamada, T., Nakano, R.: Job shop scheduling. In: Zalzala, A.M.S., Fleming, P. (eds.) Genetic Algorithms in Engineering Systems, pp. 134–160. Institution of Engineering and Technology (1997). https://doi.org/10.1049/PBCE055E_ch7

Zhang, J., Xiao, M., Gao, L., Pan, Q.: Queuing search algorithm: a novel metaheuristic algorithm for solving engineering optimization problems. Appl. Math. Model. **63**, 464–490 (2018)

Expert Systems in Information Security: A Comprehensive Exploration of Awareness Strategies Against Social Engineering Attacks

Waldson Rodrigues Cardoso[1(✉)], Admilson de Ribamar Lima Ribeiro[1], and João Marco Cardoso da Silva[2]

[1] Computing Department, Federal University of Sergipe, UFS, São Cristóvão, Brazil
{waldson.cardoso,admilson}@dcomp.ufs.br
[2] Department of Informatics, University of Minho, Braga, Portugal
joaomarco@di.uminho.pt

Abstract. This article delves into the pivotal role of expert systems in bolstering information security, with a specific emphasis on their effectiveness in awareness and training programs aimed at thwarting social engineering attacks. Employing a snowball methodology, the research expands upon seminal works, highlighting the intersection between expert systems and cybersecurity. The study identifies a gap in current understanding and aims to contribute valuable insights to the field. By analyzing five key articles as seeds, the research explores the landscape of expert systems in information security, emphasizing their potential impact on cultivating robust defenses against evolving cyber threats.

Keywords: Expert Systems · Information Security · Cybersecurity Education · Social Engineering · Snowball Methodology

1 Introduction

In today's dynamic and ever-changing digital landscape, safeguarding sensitive information against cyber threats has become an undeniable priority. Among the various challenges faced in the field of information security, social engineering attacks emerge as a particularly insidious threat, exploiting the human factor to compromise digital defenses. In this challenging context, expert systems, tools that replicate the decision-making capability of human experts, stand out in the field of artificial intelligence.

These systems are meticulously designed to address complex challenges, operating based on logical rules grounded in the "if-then" structure. Notably applied in educational decision-making, their utility extends, as demonstrated by Hwang et al. [1], to awareness and training in cybersecurity, enabling an effective educational approach.

Á. Rocha et al. (Eds.): WorldCIST 2024, LNNS 986, pp. 34–43, 2024.
https://doi.org/10.1007/978-3-031-60218-4_4

As we delve into this topic, the article will unfold in a structured manner. Initially, we will provide a brief introduction to social engineering and real-world examples of attacks, offering the essential context necessary for the subsequent analysis of the pivotal role played by expert systems. Following that, we will elucidate the snowball methodology, adopted in our research, providing insights into the approach employed for the collection and analysis of relevant literature. Subsequent sections will delve into the results from seminal works, examining the identified seeds and their contributions to understanding the intersection between expert systems and information security.

Furthermore, we will dedicate a section to the discussion, where we will deepen the analyses conducted and explore deeper connections between the findings. Finally, in the conclusion, we will summarize key findings, highlight practical implications, and point towards directions for future research. Through this structured approach, our aim is to provide a comprehensive view of the strategies adopted by expert systems to strengthen awareness and training against sophisticated social engineering attacks.

2   Social Engineering

Social engineering, as elucidated by [2], involves the manipulation, deception, and influence of individuals within an organization to comply with specific requests, ranging from disclosing sensitive information to executing tasks beneficial to the attacker. This includes simple interactions like phone calls and more complex strategies such as guiding the target to visit a website exploiting technical vulnerabilities to assume control of their computer.

In the realm of Information and Communication Technologies (ICT) security, [3] provides context for social engineering, addressing actions aimed at obtaining and exploring valuable information, including confidential data from organizations and computer systems, all by exploiting people's trust.

Defined by [4], social engineering comprises a set of techniques to manipulate people, leading them to perform actions or disclose information they wouldn't normally divulge.

Underscoring the importance of implementing and monitoring security mechanisms, coupled with specific precautions regarding electronic waste to mitigate social engineering threats, [5] emphasize the necessity of such measures. Additionally, user awareness, as highlighted by [6], is fundamental for businesses. Implementing appropriate strategies for the safe use of online social networks is crucial, aiming to prevent falling victim to social engineering attacks.

2.1   Social Engineering Attacks

Below are real examples of social engineering attacks, illustrating their diversity and sophistication:

1. Ubiquiti Networks in Hong Kong experienced a loss of approximately USD 40 million in a phishing attack, wherein attackers compromised the email account of a high-ranking employee, soliciting fraudulent payments [7].

2. During the 2016 US presidential election, hackers utilized spear phishing to create a fake email on Gmail, prompting users to change passwords. The subsequent email leak influenced the election result, favoring Donald Trump over Hillary Clinton [8] .

3. Cabarrus County in the United States fell victim to a social engineering attack resulting in a loss of USD 1.7 million. Hackers employed malicious emails, posing as suppliers, to request payments to a new bank account [9] .

4. Toyota Boshoku Corporation incurred a loss of approximately USD 37 million in a social engineering attack in Japan. Attackers persuaded a financial executive to alter bank information for an electronic transfer [10].

5. A phishing scam led to a loss of approximately USD 400,000, where a cyber-criminal posed as an assistant to a victim, soliciting payment related to real estate investments [11].

6. An 18-year-old hacker gained access to the internal network at Uber. This was a direct result of a phishing attack, and the victim, who was an employee, approved an MFA request. The attacker then looked deeper into Uber's internal systems and found a PowerShell script with admin credentials that gave him access to AWS, OneLogin, and GSuite, to name a few [12].

7. This hack was supposedly done by the same hacker who did the Uber hack. The hacker used social engineering to get into their Slack server by tricking one of the employees. He was able to grab the source code for the games GTA V and GTA VI (unreleased) and also released some of the screen recordings of the game, which seem to be a debug version [12].

These examples underscore the need for robust awareness and security strategies to effectively mitigate the diverse and sophisticated threats posed by social engineering.

Understanding the diversity and sophistication of social engineering attacks highlights the need for effective awareness and security strategies. In this context, we will now delve into the snowball methodology, a crucial approach we employ to deepen our understanding of these challenges and opportunities.

3   Snowball: Strategy

Supplementary research was imperative to elucidate the intricacies explored in this work. To achieve this, we employed the snowball methodology, a non-probabilistic sampling technique. As delineated by [13], this approach relies on pre-established inclusion criteria, offering targeted insights rather than encompassing the entire population in the sampling process.

The methodology unfolds through specific steps: the identification of pivotal documents or seeds, which conventionally mark the inception of the research line; these seeds, in turn, point to additional contributors, and the process iterates. Saturation is achieved by the repetitive emergence of themes, signaling the point where new indications cease to yield substantially novel insights or a satisfactory quantity of samples [14].

New contributors, even if indicated, undergo scrutiny to ensure alignment with defined characteristics. The snowball methodology serves exploratory purposes, aiming "[...] to understand a theme, test the feasibility of conducting a broader study, and develop the methods to be used in all subsequent studies or phases"[14].

The exploratory methodology emerges as a strategic approach to bridge the identified gap, specifically investigating how expert systems can seamlessly integrate into the context of awareness and training against social engineering threats. This endeavor aspires to make a substantial contribution to understanding the intersection between expert systems and information security practices, fostering a more holistic and informed perspective on digital protection strategies.

For this review, a search was conducted on Google Scholar from January 2 to 20, 2023, utilizing the following search terms: ("expert system") AND ("information security" OR "cybersecurity") AND ("social engineering") AND ("education" OR "awareness" OR "training"). Following predefined criteria, five works were selected as seeds. These seeds led to indications, denominated as references. The chosen works as seeds underwent rigorous selection based on the following criteria:

- Inclusion among the top 10 results returned by Google Scholar, with the condition that the article is accessible;
- Publication between 2013 and 2023, prioritizing recent works for updated results and information;
- References of the seeds adhered to the criteria of relevance to Expert Systems and Information Security, excluding books or articles from websites and blogs;

While the selected initial articles provided valuable insights into the broader landscape of expert systems and information security, it is noteworthy to mention that they did not generate additional seeds. This limitation is attributed to the thematic constraints and the specified time frame for this research (2013 to 2023). Consequently, the absence of derived seeds from these initial articles influenced the scope and breadth of the literature review, particularly in certain thematic areas. These constraints are acknowledged as inherent aspects of the research process and are explicated to provide transparency regarding the methodology's intricacies.

With the snowball methodology laying the groundwork for our investigation, we now turn our attention to the outcomes derived from the systematic review. By applying this refined methodological approach to the selected works, we can discern significant patterns and glean valuable insights into the role of expert systems in information security. Let's delve into how each seed article contributes to a broader understanding of the intersection between expert systems and digital security threats.

4   Results

In this section, we present the outcomes of the snowball review, highlighting key findings from the selected seed articles.

4.1   Summary Introduction to Results

Before delving into the specifics of each seed article, let's provide a brief overview of the main insights obtained from the selected literature.

4.2   Overview of Seed Articles

Table 1 outlines the seed articles chosen for the snowball review, each contributing unique perspectives to the intersection of expert systems and information security.

Table 1. Snowball Review Seeds

ID	Title	Reference
S1	CSAAES: An expert system for awareness of cyber security attacks	[15]
S2	Cyber attacks based on social engineering in Kenya	[16]
S3	Development of a consultative knowledge-based expert system to identify and mitigate unintentional internal threats in Ethiopian financial institutions	[17]
S4	An Effective Cybersecurity Awareness Training Model: First Defense of an Organizational Security Strategy	[18]
S5	Expert system with fuzzy logic for the protection of scientific information resources	[19]

S-1: *CSAAES: An expert system for awareness of cyber security attacks* [15] The first seed article, authored by [15], emphasizes the necessity for cyber attack protection in the context of widespread internet usage and information exposure. The presented expert system, CSAAES, focuses on raising user awareness about various attack types, symptoms, and countermeasures. While the seeds referenced by this article fall outside our specified time range, the work offers valuable insights into cyber attack awareness.

S-2: *Cyber attacks based on social engineering in Kenya* [16] Addressing cyber attacks in Kenya, the second seed by [16] explores social engineering techniques employed by cybercriminals. The article proposes an expert system-based awareness model to instill a secure mindset in users, stressing the importance of future work for model implementation and validation.

S-3: *Development of a consultative knowledge-based expert system to identify and mitigate unintentional internal threats in Ethiopian*

financial institutions [17] The third seed, authored by [17], focuses on security in financial institutions, highlighting the challenge of detecting unintentional internal threats. The article proposes a knowledge-based expert system as a practical solution to mitigate such threats in Ethiopian financial institutions.

S-4: *An Effective Cybersecurity Awareness Training Model: First Defense of an Organizational Security Strategy* [18] The fourth seed, authored by [18], emphasizes the importance of Security Education, Training, and Awareness (SETA) in organizational security strategies. The proposed model, viCyber, incorporates an expert system to develop cybersecurity curricula and enhance user understanding through real-time feedback.

S-5: *Expert system with fuzzy logic for the protection of scientific information resources* [19] The fifth seed by [19] delves into using expert systems with fuzzy logic to protect information resources in scientific libraries and corporate networks. The article underscores the benefits of expert systems in conjunction with user awareness to enhance information security.

Table 2. Comparison of articles related to the conducted research

ID	Methodology	Key Findings	Contributions
S-1	Survey on cyber security awareness programs; Integration of expert systems	Identified awareness gap; Proposed CSAAES expert system	Insights into expert systems for cyber security awareness
S-2	Case study on cyber attacks in Kenya; Analysis of social engineering techniques	Revealed social engineering tactics; Proposed expert system model	Specific approach to address social engineering threats
S-3	Exploratory applied research with quantitative approach; Rule-based knowledge representation	Effectiveness in detecting internal threats	Practical solution for internal security in financial institutions
S-4	Qualitative research using the Technology Acceptance Model (TAM); Examination of AI-based security awareness training programs	Innovative approach to cyber security education and training; Emphasized importance of SETA; Introduced viCyber model	Contribution to AI-enabled security awareness training
S-5	Analysis of the state of development of research on protecting scientific and educational databases in libraries; Application of fuzzy logic methods	Benefits of expert systems with fuzzy logic	Insights into the potential of fuzzy logic in information security

5 Discussion

The synthesis of findings from the selected seed articles sheds light on the intricate relationship between expert systems and information security, with a specific focus on awareness and training against social engineering threats. Here, we distill the key insights derived from our review:

CSAAES: An expert system for awareness of cyber security attacks [15]: Rani and Goel's work underscores the critical need for cyber attack awareness in the age of widespread internet use. The CSAAES expert system serves as a noteworthy tool for raising user awareness, emphasizing the importance of understanding various attack types and implementing effective countermeasures.

Cyber attacks based on social engineering in Kenya [16]: Obuhuma and Zivuku's exploration of social engineering techniques in the Kenyan context illuminates the vulnerabilities arising from a lack of user awareness. The proposed expert system-based awareness model stands out as a potential strategy to foster a secure mindset and combat social engineering attacks.

Development of a consultative knowledge-based expert system to identify and mitigate unintentional internal threats in Ethiopian financial institutions [17]: Adane's focus on internal threats within financial institutions highlights the challenges posed by unintentional employee actions. The proposed knowledge-based expert system offers a practical approach to identify and mitigate these threats, providing valuable insights for enhancing internal security measures.

An Effective Cybersecurity Awareness Training Model: First Defense of an Organizational Security Strategy [18]: Dash and Ansari's emphasis on Security Education, Training, and Awareness (SETA) aligns with the growing recognition of human factors in organizational security. The viCyber model, incorporating an expert system, presents an innovative approach to developing cybersecurity curricula and enhancing user understanding, showcasing the potential of expert systems in educational contexts.

Expert system with fuzzy logic for the protection of scientific information resources [19]: Normatov and Rakhmatullaev's exploration of expert systems with fuzzy logic for information resource protection emphasizes the synergy between technology and user awareness. The integration of expert systems, coupled with user education, emerges as a multifaceted strategy to enhance information security in scientific libraries and corporate networks.

In essence, these articles collectively reinforce the pivotal role of expert systems in addressing diverse facets of cybersecurity challenges. The integration of technological solutions with user education emerges as a holistic approach, acknowledging the human factor in cybersecurity. This synthesis provides a comprehensive view, setting the stage for further research and practical implementations in the dynamic field of expert systems and information security.

This synthesis not only consolidates the current understanding but also provides a comprehensive and panoramic view of the subject, thereby setting the stage for further exploration through research endeavors and practical implementations in the dynamic field of expert systems and information security.

6   Conclusion

In conclusion, the snowball review of expert systems in the context of information security, particularly focusing on awareness and training against social engineering threats, reveals a rich landscape of innovative approaches and practical solutions. The selected seed articles collectively emphasize the pivotal role of expert systems in addressing various facets of cybersecurity challenges. From raising awareness about cyber attacks to mitigating internal threats and developing effective cybersecurity training models, these articles provide valuable insights into the diverse applications of expert systems.

The synthesis of findings underscores the need for a holistic approach that combines technological solutions, such as expert systems, with user education and awareness. The effectiveness of expert systems is not only in their technical capabilities but also in their integration into broader strategies that consider the human factor in cybersecurity. The presented seed articles serve as a foundation for further research and exploration in the dynamic and evolving field of expert systems and information security.

For future research, it is essential to delve deeper into implementing the proposed expert systems, assessing their real-world effectiveness, and exploring advancements in integrating artificial intelligence with human-centric approaches for improved cybersecurity. This involves thorough evaluations of the practical deployment of expert systems, gauging their impact, and investigating innovative strategies for enhancing the collaboration between AI and human expertise. In summary, further research in these areas will contribute to refining cybersecurity measures for greater effectiveness and adaptability.

7   Recommendations for Future Research

Based on the insights gained from the snowball review, several recommendations for future research emerge:

1. **Implementation and Validation Studies:** Conduct comprehensive studies to implement and validate the proposed expert systems in real-world settings. Assess their effectiveness in diverse organizational contexts to understand their practical implications.
2. **Human-Centric Approaches:** Explore further the integration of human-centric approaches in conjunction with expert systems. Investigate how user education and awareness programs can be seamlessly integrated with expert systems to enhance overall cybersecurity resilience.
3. **Continuous Adaptation:** Given the evolving nature of cybersecurity threats, research should focus on developing expert systems that can adapt and evolve in response to emerging threats. This could involve the integration of machine learning algorithms for continuous learning and improvement.
4. **Cross-Domain Applications:** Investigate the potential cross-domain applications of expert systems in information security. Explore how the principles and frameworks developed in one domain, such as financial institutions, can be adapted and applied in other sectors.
5. **User Experience and Acceptance:** Examine the user experience and acceptance of expert systems in different organizational settings. Understand the factors that influence user trust and engagement with these systems, ultimately impacting their effectiveness.

By addressing these recommendations, future research can contribute to the ongoing advancement of expert systems in information security, fostering a more resilient and adaptive cybersecurity landscape.

Acknowledgments. Grateful for our supervisor's invaluable assistance, we extend heartfelt thanks to the Federal University of Sergipe, Brazil, and CAPES (PDPG - Strategic Partnerships in States III) for their unwavering support in completing this research. We also appreciate the contributions of the seed article authors and acknowledge the researchers and practitioners whose work forms the foundation of this snowball review.

References

1. Hijji, M., Alam, G.: Cybersecurity Awareness and Training (CAT) framework for remote working employees. Sensors (Basel). **22**(22), 8663 (2022). https://www.mdpi.com/1424-8220/22/22/8663
2. Mitnick, K.D., Simon, W.L.: The Art of Deception: Controlling the Human Element of Security. John Wiley & Sons, New Jersey (2003)
3. Hadnagy, C.: Social Engineering: The Art of Human Hacking. John Wiley & Sons, New Jersey (2010)
4. Coelho, C.F., Rasma, E.T., Morales, G.: Social engineering: a threat to the information society. Exatas & Engenharias. Higher Education Institutes of Censa. **3**(05) (2013)
5. Aramuni, J.P.C., Maia, L.C.: The impact of social engineering on information security: a management-oriented approach. AtoZ: New Pract. Inf. Knowl. **7**(1), 31–37 (2020)
6. Silva, N.B.X., Araújo, W.J.d., Azevedo, P.M.d.: Social engineering in online social networks: a case study on the exposure of personal information and the need for information security strategies. Ibero-American J. Inf. Sci. **6**(2) (2013)
7. Pinzón, J.J.S. et al.: Social engineering, the before and now of a global problem. National Open and Distance University UNAD (2015)
8. Honório, T.J.: The use of cyberattacks in elections and international relations. Mural Internacional **9**(1), 85–98 (2018)
9. Júnior, J.V.C.: Threat analysis environment for threat intelligence generation using open sources (2018)
10. Paschoal, D.A.C., Pereira, G.F.: Information security in Industry 4.0: NIST framework. Faculdade de Tecnologia de Americana (2019)
11. Dantas, A.R.P.: CYRM: Cyber Range to support the teaching of defense for students in the Information Security discipline (2022)
12. Nair, A.S.V., Achary, R.: Social Engineering Defender (SE.Def): human emotion factor based classification and defense against social engineering attacks. In: 2023 International Conference on Artificial Intelligence and Applications (ICAIA) Alliance Technology Conference (ATCON-1), Bangalore, India, pp. 1–5 (2023). https://doi.org/10.1109/ICAIA57370.2023.10169678.
13. Bickman, L., Rog, D.J.: The SAGE Handbook of Applied Social Research Methods. Sage publications, California (2008)
14. Vinuto, J.: A amostragem em bola de neve na pesquisa qualitativa: um debate em aberto. Temáticas, **22**(44), 203–220 (2014)

15. Rani, C., Goel, S.: CSAAES: an expert system for cyber security attack awareness. In: International Conference on Computing, Communication Automation, pp. 242–245 (2015). https://doi.org/10.1109/CCAA.2015.7148381
16. Obuhuma, J., Zivuku, S.: Social engineering based cyber-attacks in Kenya. In: 2020 IST-Africa Conference (IST-Africa), pp. 1–9 (2020)
17. Adane, K.: Development of advisory knowledge-based expert system to identify and mitigate unintentional insider threats in financial institutions of Ethiopia. IUP J. Comput. Sci. **14**(3), 7–23 (2020). IUP Publications
18. Dash, B., Ansari, M.F.: An Effective Cybersecurity Awareness Training Model: First Defense of an Organizational Security Strategy (2022)
19. Normatov, S., Rakhmatullaev, M.: Expert system with fuzzy logic for protecting scientific information resources. In: 2020 International Conference on Information Science and Communications Technologies (ICISCT), 1–4 (2020). https://doi.org/10.1109/ICISCT50599.2020.9351498

Multi-class Model to Predict Pain on Lower Limb Intermittent Claudication Patients

Rafael Martins[1,2]([✉]) [ID], Luís Conceição[1,2] [ID], Gustavo Corrente[3] [ID], William Xavier[3] [ID], Júlio Souza[4] [ID], Alberto Freitas[4] [ID], and Goreti Marreiros[1,2] [ID]

[1] GECAD – Research Group on Intelligent Engineering and Computing for Advanced Innovation and Development, Institute of Engineering – Polytechnic of Porto, Porto, Portugal
rgmar@isep.ipp.pt

[2] LASI – Intelligent Systems Associate Laboratory, Guimarães, Portugal

[3] Wiseware Solutions, Gafanha da Encarnação, Portugal

[4] Department of Community Medicine, Information and Health Decision Sciences (MEDCIDS), Faculty of Medicine, CINTESIS@RISE, University of Porto, Porto, Portugal

Abstract. Intermittent claudication is a vascular disease that hinders elder patients' mobility, with symptoms ranging from mild discomfort to acute pain. Based on the current literature and health professional input, the most sure-fire way to diminish these symptoms is to monitor the patient's daily walk cycle and motivate them to try to keep a steady pace despite feeling pain, the main goal being the daily increase of time the patient can walk before feeling discomfort. As such, it is of great interest to be able to predict if the patient is or will start to feel discomfort in their lower limbs, and in what specific area, based on variables such as their current speed and heart rate. By utilizing the data provided by volunteer patients throughout two months of walking tests, a few machine learning models were trained and evaluated on their success at these tasks, namely two classification tree models and two k-nearest neighbors models, one of each predicting the existence of pain and the others predicting the specific area of pain. The k-nearest neighbors models proved to be more effective at the task proposed, with a slight edge towards the pain location prediction model.

Keywords: Vascular Diseases · Intermittent Claudication · Daily Walk Monitoring · Pain prediction · Machine Learning

1 Introduction

In recent years we have witnessed a growing increase in population ageing, that is, an increase in the percentage of older population aged 60 years old and above compared to the percentage of youth. Previously an already common phenomenon in developed countries, it is now also affecting developing countries to the point the latter now host the largest portion of elderly people in the world [1]. Today they represent 7% or more of the total population in many countries, and by 2050 only 33 countries predicted not going above this threshold [2].

Á. Rocha et al. (Eds.): WorldCIST 2024, LNNS 986, pp. 44–53, 2024.
https://doi.org/10.1007/978-3-031-60218-4_5

Intermittent Claudication (IC) refers to lower extremity muscle pain, usually in the calf and less commonly in the thigh and buttock, that arises during exercise and subsides in rest, and is as such reproducible [3, 4]. This pain emerges due to insufficient oxygen delivery to meet the metabolic requirements of the muscles and is a common manifestation of peripheral arterial disease (PAD), a frequently neglected condition in the medical field that threatens not only the affected limbs but potentially the person's life as well [3, 5]. Almost a fifth of the older population has intermittent claudication, it is present in 0.6–10% of the general population, and its prevalence is expected to rise over the next years [4, 6].

As such, there is a concern in applying medical treatments to patient's suffering from intermittent claudication. Although the consequences may be severe if left unchecked, most patients can be treated with medical interventions, without the need of surgical intervention or amputation [3, 4]. Remarkably, structured walking programs to be executed regularly by the patient are reported to improve pain-free walking distance better than pharmacologic therapy alone [3].

The reported effectiveness of these structured walking plans became a major supporting factor for a health project created with the aim of remotely monitoring patient's with vascular diseases, one of them being intermittent claudication. Patients professionally diagnosed with intermittent claudication and undergoing periodical check-ups at a hospital were asked to participate in these structured walking programs, to evaluate their effectiveness. The group of volunteer patients participated in this study for approximately two months and were requested to wear two lightweight sensors each day for the longest duration they were able to (night sleep time excluded). The goal was to utilize these sensors to constantly record some of the patient's data while performing their routinely walk cycles and make it available to health professionals that need to monitor the patient.

The main sensor used was an Inertial Measurement Unit (IMU), attached to the patient's ankle through an adjustable band, capable of measuring the patient's speed, distance walked and slope angle at a given moment. An auxiliary wristband sensor, capable of recording the patient's heart rate, was also requested to be worn when possible. The sensors themselves do some data aggregation and pre-processing to obtain these variables, and then communicate with a mobile application installed on the patient's smartphone, through Bluetooth Low Energy (BLE) technology, to transmit the data. In addition to the two sensors, the patients were asked to indicate on the mobile application whenever they started and stopped feeling pain during their walks. This could be done by selecting the respective area of the leg where they felt pain, in a dedicated lower limb map shown in the application, and deselecting it once the pain subsided.

All data collected from the sensors and the patient's input was automatically and periodically transferred from the mobile application to an HL7 FHIR-compliant server and database, to ensure standardization and data privacy, and was then made available to health professionals through different dashboards.

Finally, the data collected through the two trial months was exported from the database and used to train and evaluate different machine learning models that utilize the patient's speed, distance, heart rate, and slope angle, to predict if they have pain and estimate its location.

An envisioned practical application for this model would be its integration into a system which alerts health professionals when the patient feels (or would be feeling) pain. This allows them to study the positive or negative impact of the continuous walking treatment on a patient, by comparing the pain predictions trained on previous reports by the patients with the actual current reported pain moments. Another potential use for the model is for it to be integrated into a mobile application for the patient that notifies them that it is predicted they will feel pain soon, as a way to reassure the patients that the pain is normal and that they should continue walking or motivate the patient by showing them their progress in the distance they can walk while being pain free.

2 Case Study

Following the reported success of structured walking plans for the treatment of intermittent claudication, patients were recruited to measure health data during their daily walk cycles, for a period of approximately two months. The vital data recorded from sensors was aggregated per second, and it consisted of the patient's current speed (m/s), total distance walked thus far (m), current heart rate (bpm), and current slope angle (°). Simultaneously, during their walk cycles the patients signaled in a mobile application every time started and stopped feeling pain, and in what locations.

Synchronizing the recorded vital data with the periods of pain reported by the patient resulted in a labeled dataset with the vital data as inputs, and one or a combination of pain locations as the output in each entry.

2.1 Feature Extraction and Correlation

After processing the two months of data for relevant feature extraction, a correlation matrix was calculated to analyze possible linear correlations between the sensor-measured variables (i.e., the patient's current walking speed, current total walking distance for that day, current heart rate, and current slope angle) and the presence or not of

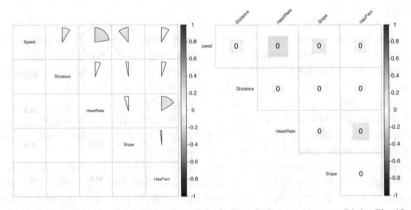

Fig. 1. Correlation Matrix of the data collected. Left: Correlation percentages; Right: Significance levels (p-values) of the correlations

pain reported by the patient. The exploration of these correlations improved our understanding of intermittent claudication and helped selecting the relevant variables to utilize during the machine learning models' training (Fig. 1).

Every result was statistically significant, and so from the constructed correlation matrix some moderate linear correlations could be observed, according to the Chan YH correlation interpretation scale, commonly used in the field of medicine[7, 8]. These correlations show some noteworthy patterns:

- A higher walking speed, and to a less extent a higher total distance walked, can raise the patient's heart rate levels.
- An increased slope can somewhat reduce the patient's walking speed.
- Increased speed, distance, and especially heart rate contribute to a higher probability of the patient having pain.

The first two patterns correspond to normal expectations, helping to confirm the validity of the data, while the third one was the main objective of this analysis. From this, walking speed, walking distance and heart rate were deemed as important variables for training the pain prediction machine learning models. Slope angles, while having been classified as a poor correlation, were in the end also utilized in the model training due to their influence on the speed and distance of patients.

2.2 Pain Prediction Models

In order to study the use of machine learning models for pain prediction in lower limb intermittent claudication, two different types of algorithms were utilized: classification tree algorithms, and k-nearest neighbors algorithms. In both cases two models were trained: a binary classification model, and a multiclass classification model.

The binary models, designated as "pain presence" prediction models, only predict if the patient is feeling pain or not at a given moment, having as outputs "True" or "False". The multiclass models, designated as "pain location" prediction models, go further by trying to predict not only if the patient is feeling pain, but also its location in the patient's lower limbs. Since the patient can report having pain in multiple locations simultaneously, the output is either "No Pain", or it is a combination of one or more of the labels "Calf", "Thigh" or "Shin", followed by "-Left" or "-Right" to distinguish the affected leg. In other words, one possible output for where the patient is feeling pain may be "Calf-Left & Thigh-Right".

All four models utilize as inputs the patient variables analyzed in the feature extraction phase: current speed, total walked distance (in the current day), current heart rate, and current slope angle.

A noteworthy factor, however, is that, since the data collected spans the patient's condition during each day, the data ended up heavily imbalanced since the majority of instants captured reported "No Pain". In order to mitigate the issues that this imbalance causes in the training of prediction models, some undersampling techniques were applied to partially reduce the rate of data with no pain reported, while still retaining some of this imbalance so as to not disregard the context of the data.

Classification Tree Models. Classification Trees are a subset of decision tree learning models where the target variable takes a discrete set of values. They are a supervised machine learning approach that receive training data where the target variable is known and construct a tree composed of decisions branch nodes to take to reach a target variable's possible classified label, represented as leaf nodes.

To train these models, 70% of the data obtained from volunteer patient trials was used as train data, leaving the other 30% for subsequent model evaluation.

K-Nearest Neighbors Models. The K-Nearest Neighbors algorithm (KNN) is a non-parametric supervised machine learning strategy that classifies each new data point based on the classified labels and distances of the K nearest training data points. Simply put, each new data point is grouped together with the label of the majority of its neighbors.

Since this algorithm is based on distances, it is reported that normalizing the data can increase the model's accuracy dramatically [9]. After data normalization, the same set of train data as in the classification tree algorithm (70% of the total data) was used, and a value of $K = 7$ was chosen for the number of near neighbors to consider.

3   Results

The four machine learning models were all trained with the same 70% subset of the patient's data and evaluated with the same reserved 30% so that not only each model's accuracy and effectiveness could be evaluated, but also the comparative performance between the two different algorithms.

3.1   Classification Tree Models

The Classification Tree algorithms resulted in models with similar decisions between variables along the decision tree calculation, the only difference being that in the pain presence model the only possible classifications were "True" if the patient is predicted to feel pain, or "False" otherwise, while the pain location model had more detailed outputs depending on the area of pain. A representation of the decision tree calculated by the pain location prediction model is presented in Fig. 2.

By inputting the previously reserved 30% of patient data, it was possible to evaluate the accuracy of the models by comparing the predicted pain existence/location with the reference values, the real values from the test data. In each model, this comparison was done through a confusion matrix, one presented in Fig. 3, which in turn allowed for detailed statistical analysis of the models' performance.

Both learning models reported an accuracy rate of 98.9% percent. However, the accuracy metric is not reliable alone since it is influenced by imbalances present in the data collected, in this case the significantly larger quantity of samples with no pain reported (represented by either "False" or "No Pain" depending on the model).

As such, an important test to determine is if the accuracy rate is significantly higher than the No Information Rate (NIR), that is, the rate of data in the majority class, which both models reported to be 98.2%. The p-value obtained from a binomial hypothesis

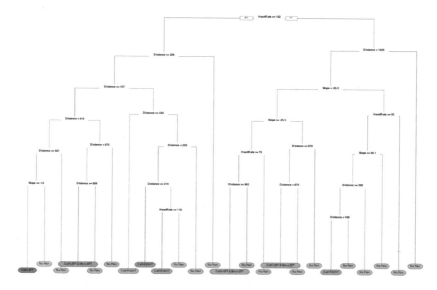

Fig. 2. Representation of the classification tree pain location prediction model

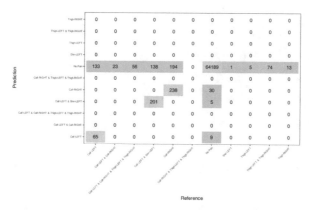

Fig. 3. Confusion Matrix for Statistic Evaluation of the Classification Tree pain location prediction model.

test was vastly inferior to 0.05, meaning the accuracy percentage is at least somewhat statistically significant despite the data imbalance.

To better evaluate these models, the Cohen's Kappa value was calculated, a statistic similar to the accuracy ratio that also considers the probability of the prediction having been selected at random due to data imbalances. Both models reported a low Cohen's Kappa value of approximately 57%. In fact, the sensitivities reported by these algorithms, a statistic that represents the capacity of the models to correctly predicting pain and its location, compared to the number of false positives, were all under 55%.

3.2 7-Nearest Neighbors Models

To train a K-Nearest Neighbors model, it is necessary to specify a value for K, the number of neighbors the algorithm will use to predict a classification for each new data point, and this value can affect the accuracy and performance of the model. By training some models with different K values, the two 7-Nearest Neighbors models (pain presence and pain location predictions) reported higher levels of efficacy.

In a similar fashion to the Classification Tree models, both algorithms resulted in similar models, with the only difference again being the more detailed labels that the pain location prediction model gives. A partial representation of the pain location model is presented in Fig. 4.

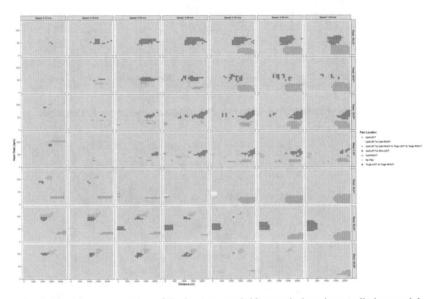

Fig. 4. Partial representation of the k-nearest neighbors pain location prediction model

Due to the nature of this algorithm and of the dataset collected from the patient trials, the pain predictions are calculated by measuring the distances to neighbor points in four dimensions (speed, distance, heart rate and slope), making a total representation of the algorithm challenging. In Fig. 4, each smaller graph plots how the changes in distance and heart rate affect the algorithms prediction of the pain's location, for a specific constant speed and slope angle. Then, the full grid of graphs shows how changes in speed (x axis), and slope angle (y axis) also affect the prediction. Note that graphs where the speed value was inferior to 0.7 m/s were truncated, since the model always predicted "No Pain", regardless of variations in the other variables.

For model evaluation, the methodology established for the Classification Tree models was used. For each model, a confusion matrix, presented in Fig. 5, was generated to compare the predicted pain existence/location with the reference values, which in turn allowed for a detailed statistical analysis of the models' performance.

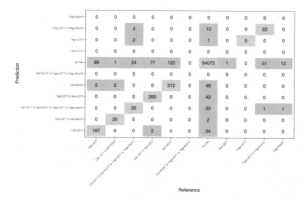

Fig. 5. Confusion Matrix for Statistic Evaluation of the 7-Nearest Neighbors pain location prediction model.

Both models reported an accuracy of 99.2%, slightly higher than the classification tree models, and the calculation of the binomial hypothesis test determined that the accuracy rate is statistically significantly higher than the No Information Rate (NIR), showing that this accuracy rate is relevant despite the data imbalance.

To take into account the probability of the prediction being chosen at random due to data imbalances, the Cohen's Kappa value was calculated an in both models it reported a significantly higher value of approximately 73% compared to the classification trees. Finally, these models also show a higher capacity of classifying true positives and avoiding false positives when compared to the classification tree models, with the pain presence prediction model reporting a sensitivity of 67%, and the pain location model reporting sensitivities above 75% for the pain location combinations that include the calves, but sensitivities of 30% for less reported pain locations.

4 Discussions

This study involved analyzing approximately two months data collected from patients professionally diagnosed with intermittent claudication. This data was then used as inputs to train and evaluate models that try to predict if the patient is in pain, and the location of that pain in their lower limbs. This section discusses the most important observations and conclusions obtained from the results previously presented.

The first step of the study consisted in creating a correlation matrix to identify possible linear correlations between the data variables. As expected from literature and professional advice, the measured variables (speed, distance, heart rate, and slope angle) appear to be correlated with the patient's pain, and therefore useful to try to predict it. However, the correlations observed were not as strong as expected, with the main probable cause being the strong imbalance present in the data towards the patient feeling no pain. Medically, this imbalance is expected and desirable, as the patient's should not constantly be reporting pain during their daily walk cycles. From a data science perspective, this imbalance did affect the training of the models and consequently their prediction capabilities. This can be mitigated in future work by trying to recruit patients that more frequently report feeling pain, and by applying heavier undersampling techniques to further reduce the rate of data with no pain reported.

For each algorithm used, two models were trained: a binary model that only predicts the existence or absence of pain, and a multiclass model that predicts the pain's location. The idea was that the binary models alone were useful to alert health professionals or an automatic system of the patient's, while the purpose of the pain location model was to give more detailed alerts, but at the cost of not being as precise. Contrary to these expectations however, both models (from the same algorithm type) reported the same accuracy and performance, with even a slight edge to the pain location prediction models, rendering the binary models mostly obsolete.

Comparing the performance of the two machine learning algorithms, the 7-Nearest Neighbors models were observed to be more accurate and more sensitive to identifying true positives. Both types of models showed that the patient's heart rate was an important factor to predicting pain, corresponding to the expectations formed from the correlation matrix analysis, but the 7-Nearest Neighbors models also gave higher importance than expected from this first analysis to the patient's current speed, predicting that any speed below 0.7 m/s usually leads to no pain, regardless of the other variables' variation. Yet another noteworthy observation was that, while most pains were reported by the patients to be located in the calves, as is described in literature regarding intermittent claudication [4], the 7-Nearest Neighbor model represented in Fig. 4 predicts that a higher slope angle causes more pain in the thighs, suggesting that the patients may be exerting too much force on their thighs when walking uphill.

5   Conclusions and Future Work

Recent years have shown a growing increase in population ageing. Today, people aged 60 years old and above comprise 7% or more of the total population in many countries, and by 2050 only 33 countries are predicted to not go above this threshold [1, 2]. Intermittent Claudication (IC) is a vascular disease that affects almost a fifth of the older population, and is present in 0.6–10% of the general population, with its prevalence being expected to rise even further in the next years [4, 6]. Patients diagnosed with intermittent claudication report feeling lower extremity muscle pain, usually in the calf, thigh, shin and buttock, that arise during exercise and subsides in rest [3, 4]. This pain emerges due to insufficient oxygen delivery to meet the metabolic requirements of the muscles, and is a frequently neglected condition in the medical field that threatens not only the affected limbs but potentially the person's life as well [3, 5]. As such, there is a concern in applying medical treatments to patient's suffering from intermittent claudication, with structured walking programs to be executed regularly by the patient being reported as one of the better ways to improve the pain-free walking distance [3].

With this in mind, this study utilized data collected from patients for approximately two months to train and evaluate a binary model that predicts pain presence or absence and a multiclass model that predicts pain location, each through two different machine learning algorithms: Classification Trees, and K-Nearest Neighbors.

The 7-Nearest Neighbors algorithms were observed to be more accurate than the Classification Trees, and exceeding expectations the multiclass pain location model reported slightly better accuracy than the binary pain presence model. However, all models were negatively influenced by some expected imbalances in the data towards no pain reports, which is why a possible future work may be to repeat a similar study with recruited patients who report having pain more frequently, and also apply further undersampling techniques to the high rate of no pain entries in the data.

Regardless, the 7-Nearest Neighbors pain location prediction model proved to be effective at the task proposed. In future work, this model has the potential to be integrated into a system for health professionals, which can alert them when the patient feels pain, and allow them to study the positive or negative impact of the treatment on the patient by comparing the pain predictions trained on previous reports by the patients with the actual current reported pain moments. The model can also be integrated into a mobile application for the patient that notifies the patient that it is predicted they will feel pain soon, as a way to reassure the patients that the pain is normal and that they should continue walking or motivate the patient by showing them their progress in the distance they can walk while being pain free.

Acknowledgments. This research work was developed under the project RM4Health.
(ITEA-2021–21022-RM4Health), funded by the European Regional Development.
Fund (ERDF) within the project number COMPETE2030-FEDER-00391100, and by National Funds through the Portuguese FCT — Fundação para a Ciência e a Tecnologia under the R&D Units Project Scope, UIDB/00760/2020 (https://doi.org/10.54499/UIDB/00760/2020), UIDP/00760/2020 (https://doi.org/10.54499/UIDP/00760/2020).

References

1. Ismail, Z., Ahmad, W.I.W., Hamjah, S.H., Astina, I.K.: The impact of population ageing: a review. Iran. J. Public Health **50**(12), 2451–2460 (2021). https://doi.org/10.18502/ijph.v50i12.7927
2. He, W., Goodkind, D., Kowal, P.: An Aging World: 2015 (2016). https://doi.org/10.13140/RG.2.1.1088.9362
3. Patel, S.K., Surowiec, S.M.: Intermittent claudication. In: StatPearls. StatPearls Publishing, Treasure Island (FL) (2023). http://www.ncbi.nlm.nih.gov/books/NBK430778/. Accessed 15 Nov 2023
4. Cassar, K.: Intermittent claudication. BMJ **333**(7576), 1002–1005 (2006). https://doi.org/10.1136/bmj.39001.562813.DE
5. Diehm, C., Kareem, S., Lawall, H.: Epidemiology of peripheral arterial disease. VASA Z. Gefasskrankheiten **33**(4), 183–189 (2004). https://doi.org/10.1024/0301-1526.33.4.183
6. Garcia, L.A.: Epidemiology and pathophysiology of lower extremity peripheral arterial disease. J. Endovasc. Ther.Endovasc. Ther. **13**(2_suppl), II-3–II−9 (2016). https://doi.org/10.1177/15266028060130S204
7. Akoglu, H.: User's guide to correlation coefficients. Turk. J. Emerg. Med. **18**(3), 91–93 (2018). https://doi.org/10.1016/j.tjem.2018.08.001
8. Chan, Y.H.: Biostatistics 104: correlational analysis. Singapore Med. J. **44**(12), 614–619 (2003)
9. Hastie, T., Tibshirani, R., Friedman, J.: Prototype methods and nearest-neighbors. In: Hastie, T., Tibshirani, R., Friedman, J. (eds.) The Elements of Statistical Learning, pp. 459–483. Springer New York, New York, NY (2009). https://doi.org/10.1007/978-0-387-84858-7_13

Collaborative Filtering Recommendation Systems Based on Deep Learning: An Experimental Study

Eddy Pardo, Priscila Valdiviezo-Diaz[✉], Luis Barba-Guaman, and Janneth Chicaiza

Department of Computer Science, Universidad Técnica Particular de Loja, San Cayetano Alto, Loja 1101608, Ecuador
{eapardox,pmvaldiviezo,lrbarba,jachicaiza}@utpl.edu.ec

Abstract. Recommender systems allow users to filter relevant information, helping users discover content and products that fit their preferences and interests. Collaborative filtering is one of the most widely used approaches in recommender systems, which uses historical user data for the recommendation. Nowadays, researchers are exploring new ways to make recommendations, using deep network architectures that have a major impact on some areas. This paper explores collaborative filtering recommendation systems based on deep learning, focusing on the experimental evaluation of algorithms most used in these systems. In our experimental study, we evaluate the performance of Autoencoder and Neural Collaborative Filtering models on representative datasets. As the main result, we found that the Autoencoder model outperformed Neural Collaborative Filtering in terms of prediction, suggesting its usefulness by providing more precise recommendations for users.

Keywords: Collaborative filtering · Deep learning · Neuronal Collaborative filtering · Recommender systems

1 Introduction

Recommender systems (RS) have become very important in recent years, due to the possibilities they offer to filter relevant information for the user [18]. These systems provide users with a personalized and enriching experience by selecting relevant content from a vast universe of options. Among the prominent approaches, collaborative filtering has proven to be a valuable technique by leveraging the collective information of users to generate recommendations.

Currently, new ways to take better advantage of the amount of data available are being investigated, with the use of deep network architectures, which are generating a great impact in some areas of knowledge. These new architectures can be used to improve the accuracy and generalization capacity of recommender systems.

Á. Rocha et al. (Eds.): WorldCIST 2024, LNNS 986, pp. 54–63, 2024.
https://doi.org/10.1007/978-3-031-60218-4_6

Deep learning-based RS has gained significant attention by overcoming the obstacles of conventional models and achieving high quality recommendations. These systems show significant advantages in improving the quality of recommendations [1] attempting to solve the user cold-start problem by integrating linear matrix factorization features with the multilayer perceptron's nonlinear features, and employing implicit trust information to generate effective recommendations.

Some studies on deep learning-based recommender systems have been realized, for example, reference [20] provides a comprehensive review of the related research contents of deep learning-based recommender systems. In [6] present a study on product recommendation systems based on deep learning and collaborative filtering. Chen et al.[8] carry out an overview of the representative heterogeneous one-class collaborative filtering methods from the perspective of deep learning-based methods. Likewise, reference [2] presents a comparative study on deep learning algorithms for the cold-start problem in recommendation systems.

The objective of our paper is to provide an understanding of the application of Deep learning techniques in product recommendation, with special emphasis on collaborative filtering recommendation systems. Furthermore, through an experimental study, we seek to understand the comparative performance of the deep learning models in representative datasets.

To contribute to the advancement of research in deep learning-based recommender systems, this paper makes the following contributions:

- A review of the state-of-the-art collaborative filtering recommendation systems based on deep learning and a discussion of the challenges of these systems.
- An experimental study evaluating the performance of deep learning algorithms on benchmark datasets, comparing their performance based on conventional metrics.

The structure of the paper is organized as follows: Sect. 2 shows a description of the deep learning-based recommender system. Section 3 includes the experimental results to measure the quality of the prediction and recommendation of the deep learning algorithms. Section 4 encloses the conclusions of this contribution and future work.

2   Deep Learning-Based Recommender System

Deep learning-based recommender systems can be divided into two broad types: integration model and neural network model [20]. Namely, whether it incorporates conventional recommendation models in conjunction with deep learning methodologies or exclusively relies upon deep learning techniques.

In the literature review, it was found that a large number of works integrate conventional recommendation models with deep learning technologies. In this work, we will present details of deep learning-based collaborative filtering recommender systems.

2.1 Deep Learning-Based Collaborative Filtering Recommender Systems

Collaborative filtering (CF) is the most commonly used approach in RS, which can be of two types, user-based and item-based. This approach uses information about users' interests or preferences to make automated predictions [11]. CF often faces the data sparsity problem, which refers to the recommendation process becoming inefficient; and the cold-start problem where the system cannot offer a recommendation because it has not yet gathered sufficient user or item information. Some works have been carried out to address this type of problems, for example, [5] proposes a deep nonlinear non-negative matrix factorization (DNNMF) technique to address the data sparsity problem. In [3], a neural recommendation model is proposed based on non-independent and identically distributed for CF by incorporating explicit and implicit coupling interaction. The authors use the coupling interactions between the users and items to overcome the recommender system's cold-start and sparsity issues. Reference [19] proposes a hybrid approach using a deep neural network to estimate CF vectors to handle the cold-start problem in RS.

The utilization of deep learning in CF recommendation systems offers advantages such as effective handling of sparse data, dynamic adaptability to evolving preferences, scalability, and higher recommendation precision.

A more detailed analysis of some works on collaborative filtering recommendation systems based on deep learning is presented in Table 1.

The analyzed works provide valuable insights into deep learning-based recommender systems, such as the diverse applications (e-commerce, social networks, entertainment, etc.) and the challenges of these systems. Furthermore, Table 1 shows that recommendation approaches can be applied to both machine learning and deep learning algorithms for the recommendation. We can see that different metrics have been used to assess the quality of the predictions and recommendations, for example, [9] measures the performance of the algorithm using Amazon datasets and metrics such as RMSE, MAE, and NMAE to evaluate the quality of the predictions; and, for the evaluation of the recommendation they use metrics such as MRR, MAP, and NDCG.

In Table 1, we can see that to evaluate recommendation quality the Normalized Discounted Cumulative Gain metric (NDCG) is the most used, followed by the Recall metric used to know how many positive values are correctly classified, and Hit Radius (HR) that helps to measure the effectiveness with respect to the list of recommendations. To measure the prediction quality the Root Mean Squared Error metric (RMSE) is the most used.

2.2 Deep Learning Models

Some deep learning models can be used in collaborative filtering recommendation systems. In this section, we describe commonly used deep learning models.

- Autoencoder (AE) is an unsupervised model attempting to reconstruct its input data in the output layer [20]. It uses a nonlinear hidden layer encoder

Table 1. Summary of the works analyzed.

Reference	Objective	Deep learning algorithm	Metrics
[9]	This paper evaluates a recommendation approach that integrates sentiment analysis into collaborative filtering methods based on an adaptive architecture, which includes deep learning models.	CLSTM, LCNN	RMSE, MAE, NMAE, MRR, MAP, NDCG
[25]	This paper develops a hybrid RS utilizing a deep autoencoder network. The proposed approach employs collaborative and content-based filtering, as well as users' social influence	SRDNet, AutoRec, VB-CF	MAE, RMSE
[17]	This paper proposes a Path-based Deep Network incorporating personalization and diversity to enhance matching performance.	PDN	HR, NDCG
[13]	Authors explore the traditional ways of making a recommender engine and then evaluate the use of deep learning Autoencoder on a large dataset.	AE	RMSE, MAE, Accuracy, Recall
[26]	This paper introduces a hybrid collaborative filtering framework that applies Matrix factorization and CNN in a parallel manner to learn knowledge from implicit feedback data.	CNN	HR, NDCG
[15]	This paper proposes a visual recommendation system based on Deep Visual Ensemble similarity metric using Convolutional Autoencoder neural network for classification.	CNN, CAE	AUC, RMSE, Recall
[27]	This research designs a deep RS that discovers user preferences and provides the reasons behind the recommendations it makes by examining clues such as user reviews and item descriptions.	CNN	Precision, Recall, F1 score
[4]	This paper shows a multi-criteria collaborative filtering algorithm based on autoencoders in order to nonlinearly represent relations among users in terms of multi-criteria preferences.	AE-MCCF	MAE, RMSE
[23]	This paper proposes a deep learning based content-collaborative methodology for personalized size and fit recommendation.	NCF, SFnet	Accuracy, AUC
[7]	This research proposes a deep neural network architecture that uses ID embeddings and auxiliary information such as features of job postings and candidates.	NeuMF, DNN,	NDCG, HR

and decoder blocks and eventual reconstruction of original data [24]. The algorithm inputs are embedded in vectors and fed into a deep neural network, as the number of hidden layers increases, the model learns more complex higher order data patterns at the cost of increased computational time.

- Recurrent neural networks (RNN) are effective at modeling sequential data, which can be useful for recommender systems that consider the evolution of user behavior over time. In this structure of neural network, there are loops and memories to remember former computations [20].
- Convolutional neural network (CNN) is a special kind of feedforward neural network with convolution layers and pooling operations [16]. These networks are useful for learning spatial representations in data as images and can be applied in recommendation systems to process visual information associated with products.

These models combined with collaborative filtering are part of the model-based collaborative filtering methods, which have also received other names such as, for example, Autoencoder-based Collaborative Filtering (ACF) is an autoencoder-based collaborative recommendation model. Instead of using the original partial observed vectors, ACF decomposes them by integer ratings [22]. Likewise, the RNN-based collaborative filtering is to use RNN to model the effect of user historical sequence behavior on the user current behavior, then recommends items for the user predicts user's behavior [28].

2.3 Challenges of Collaborative Filtering Recommender Systems Based on Deep Learning

Nowadays, the volume of information on the web is growing due to the increase of new visitors and articles available on the Internet, which brings new challenges for deep learning-based RS, such as training large-scale deep learning models. Moreover, it is necessary to study the recommendation of multiple user behaviors that cannot form a full-order cascading relationship, so these behaviors not only contain normal interactions between users and articles but may also include social interactions between users, such as sharing, following, etc. So it becomes necessary to investigate how to integrate these heterogeneous types of dynamic user behaviors into a unified recommendation framework [12].

According to [10], one of the most important challenges to consider is the incorporation of additional user (profiles, behavior) and item information to improve the performance of many RS models.

Deep learning models, especially more complex ones like deep neural networks, tend to be black boxes that are difficult to interpret. In some domains, where explainability is essential to building trust, this may be a concern. In addition, intensive use of high-capacity computing to train and maintain deep learning models may be necessary. This can be a barrier for smaller companies or those with infrastructure limitations.

Although recommender systems researchers have done several studies to address the problems of cold-start, sparsity, and scalability, new solutions to these problems can still be formulated to contribute to research in the context of recommender systems.

Finally, there are still some challenges of recommender systems, which need to be addressed from the perspective of deep learning as the Shilling attack problem, that is, when a malicious user fakes his identity and enters the system to give false item ratings, y Synonymy problem when similar items have different entries or names, in the system [21].

3   Results

In this section, we include the results obtained from the experiments with two selected algorithms: Autoencoder (AE) and Neural collaborative filtering (NCF) which is a multilayer representation algorithm for modeling a user-element interaction [14]. In order to analyze the behavior of algorithms the following quality measures have been used: Root Mean Square Error (RMSE) to measure the quality of the predictions; Precision and Recall to measure the quality of recommendations.

3.1   Experimentl Setup

For the execution of the experiments, we utilized two datasets: The MovieLens-1M[1] dataset and the Books dataset[2] available on the Kaggle repository. MovieLens-1M contains 1,000,209 ratings from 6,040 users to 3,076 items on a 5-star scale. The books dataset contains 231,829 ratings from 8,902 users to 9,999 items on a 5-star scale.

Experiments are performed with different values in the configuration hyperparameters for each algorithm. These have been selected in order to maximize the accuracy of algorithms for all quality measures utilized.

The initial dataset is split into 80% training and 20% test users: training users will be used to train both algorithms and test users will be used to measure the performance of models. The experiments have been carried out using Python. We use Keras, an open-source deep learning framework available for TensorFlow.

3.2   Experimental Results

Table 2 contains the RMSE values and the execution time for both MovieLens-1M and Books dataset. It is observed that AE provides more accurate predictions than the NCF algorithm with a regularization parameter of $1e-6$ for MovieLens-1M and a regularization parameter of $1e-5$ for the Books dataset. Regarding the computational performance in terms of execution time from Table 2, we can see how the execution time increases with NCF for both datasets. Therefore, the AE algorithm is significantly better than the NCF algorithm for the two datasets (MovieLens-1M and Books).

---

[1] https://grouplens.org/datasets/movielens/1m/.
[2] https://www.kaggle.com/code/philippsp/book-recommender-collaborative-filtering-shiny/.

Table 2. Root Mean Square Error (RMSE) and the execution time of each recommendation algorithm for both MovieLens and Books datasets.

	MovieLens-1M		Books dataset	
Algorithm	RMSE	Time (minutes)	RMSE	Time (minutes)
AE	0.1526	11.22	0.2727	16.22
NCF	0.4587	53.93	0.4302	59.85

In addition, comparing the results with other state-of-the-art works with Movielens-1M, we can see that the RMSE value obtained in our experiment is better than the one obtained by reference [22] of 0.831; this difference is due to the configuration of regularization parameters.

To evaluate the quality of the recommendations, we selected the algorithm with the best results in the RMSE metric. Figure 1 contains the Precision and Recall values for the MovieLens dataset with the AE algorithm. Precision and Recall have been tested using the relevance threshold at value 4 to discriminate if a recommendation is relevant or not. As we can see for a top 10 recommendations, the precision value decreases as the number of recommendations increases. On the other hand, the recall value increases as the number of recommendations increases.

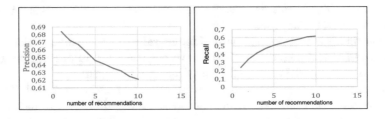

Fig. 1. Precision and recall for MovieLens-1M dataset.

In terms of Precision, we can observe that 62% of the recommendations are relevant to the user. On the other hand, the Recall indicates that 60% of the relevant movies were recommended to the user.

Figure 2 contains the Precision and Recall values with the AE algorithm for the Books dataset, where the Precision value decreases as the number of recommendations increases. On the contrary, the Recall value increases as the number of recommendations increases. That is, for the top 10 recommendations, 65% of the recommendations are relevant to the user, and 60% of the relevant books were recommended within the top 10 books.

From Table 2 and Figs. 1 and 2, we can conclude that the AE algorithm is better in RMSE in the two datasets, and in the precision metric when the number of recommendations decreases.

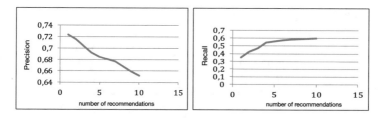

Fig. 2. Precision and recall for Books dataset.

Besides, when making a comparison between the Autoencoder model and the Neural Collaborative Filtering model in terms of performance and accuracy we can see depends on various factors, including the dataset, hyperparameter tuning, and specific implementation details.

4 Conclusions and Future Work

This paper introduces a study on collaborative filtering recommendation systems based on deep learning and an experimental evaluation focused on movie and book recommendations, which has allowed us to understand the effectiveness of deep learning algorithms as well as obtain substantial contributions to the understanding of this field of fast growth.

Our study identifies the work carried out on recommendation systems based on deep learning, providing a vision of the deep learning models commonly used in the literature and the challenges of these systems. In addition, we have analyzed the state-of-the-art collaborative filtering systems based on deep learning, highlighting the algorithms and metrics used in these works.

The experimental study includes the comparison of the effectiveness of the Autoencoder and Neural Collaborative Filtering (NCF) models on two representative data sets. The results reveal that the autoencoder model presented better prediction performance, outperforming the NCF model. These findings point to the promising utility of autoencoder models in the context of collaborative filtering recommendation systems based on deep learning.

From the study conducted, it was determined that most of the works focused on deep learning-based recommender systems are focused on alleviating the main problems that arise in recommender systems such as cold-start, scalability, and data sparsity, so it is necessary to continue with the line of research to solve other problems that arise in recommender systems such as Shilling attack and synonymy problems.

The following topics are proposed as future work: a) To explore the robustness and scalability of the models evaluated on larger and more diverse data sets; b) To extend the experiments with other deep learning models and compare the results with the models evaluated in this study.

References

1. Ahmed, A., Saleem, K., Khalid, O., Rashid, U.: On deep neural network for trust aware cross domain recommendations in e-commerce. Expert Syst. Appl. **174**(C), 114757 (Jul 2021). https://doi.org/10.1016/j.eswa.2021.114757
2. Alabdulrahman, R., Viktor, H., Paquet, E.: Active learning and deep learning for the cold-start problem in recommendation system: a comparative study. Commun. Comput. Inform. Sci. **1297**, 24–53 (2020)
3. Aljunid, M.F., Huchaiah, M.D.: Integratecf: integrating explicit and implicit feedback based on deep learning collaborative filtering algorithm. Expert Syst. Appl. **207**, 117933 (2022). https://doi.org/10.1016/j.eswa.2022.117933
4. Batmaz, Z., Kaleli, C.: AE-MCCF: An autoencoder-based multi-criteria recommendation algorithm. Arabian J. Sci. Eng. **44**, 9235–9247 (05 2019)
5. Behera, G., Nain, N.: DeepNNMF: deep nonlinear non-negative matrix factorization to address sparsity problem of collaborative recommender system. Int. J. Inform. Technol. (Singapore) **14**(7), 3637–3645 (2022)
6. Bhagat, M.D., Chatur, P.N.: A study on product recommendation system based on deep learning and collaborative filtering (2023), cited by: 1
7. Çakır, M., Öğüdücü, ŞG., Tugay, R.: A deep hybrid model for recommendation systems. In: Alviano, M., Greco, G., Scarcello, F. (eds.) AI*IA 2019 - Advances in Artificial Intelligence, pp. 321–335. Springer International Publishing, Cham (2019)
8. Chen, X., Li, L., Pan, W., Ming, Z.: A survey on heterogeneous one-class collaborative filtering. ACM Trans. Inf. Syst. **38**(4), 1–54 (Aug 2020). https://doi.org/10.1145/3402521
9. Dang, C., Moreno GarcÃa, M., De La Prieta, F.: An approach to integrating sentiment analysis into recommender systems. Sensors **21**(16), 5666 (08 2021). https://doi.org/10.3390/s21165666
10. Da'u, A., Salim, N.: Recommendation system based on deep learning methods: a systematic review and new directions. Artif. Intell. Rev. **53**(4), 2709–2748 (2019). https://doi.org/10.1007/s10462-019-09744-1
11. Devika, R., Subramaniyaswamy, V.: A novel model for hospital recommender system using hybrid filtering and big data techniques. In: 2018 2nd International Conference on I-SMAC (IoT in Social, Mobile, Analytics and Cloud) (I-SMAC)I-SMAC (IoT in Social, Mobile, Analytics and Cloud) (I-SMAC), pp. 267–271 (2018)
12. Gao, C., et al.: Learning to recommend with multiple cascading behaviors. IEEE Trans. Knowl. Data Eng. **33**, 2588–2601 (12 2019)
13. Guha, R.: Improving the performance of an artificial intelligence recommendation engine with deep learning neural nets. In: 2021 6th International Conference for Convergence in Technology (I2CT), pp. 1–7 (2021). https://doi.org/10.1109/I2CT51068.2021.9417936
14. He, X., Liao, L., Zhang, H., Nie, L., Hu, X., Chua, T.S.: Neural collaborative filtering. In: Proceedings of the 26th International Conference on World Wide Web, pp. 173–182. WWW '17, International World Wide Web Conferences Steering Committee, Republic and Canton of Geneva, CHE (2017)
15. Hiriyannaiah, S., Siddesh, G., Srinivasa, K.: Deep visual ensemble similarity (dvesm) approach for visually aware recommendation and search in smart community. J. King Saud Univ. - Comput. Inform. Sci. **34**(6, Part A), 2562–2573 (2022)

16. Krizhevsky, A., Sutskever, I., Hinton, G.E.: Imagenet classification with deep convolutional neural networks. In: Pereira, F., Burges, C., Bottou, L., Weinberger, K. (eds.) Advances in Neural Information Processing Systems. vol. 25. Curran Associates, Inc. (2012)
17. Li, H., et al.: Path-based deep network for candidate item matching in recommenders. In: Proceedings of the 44th International ACM SIGIR Conference on Research and Development in Information Retrieval, pp. 1493–1502. SIGIR '21, Association for Computing Machinery, New York, NY, USA (2021)
18. Li, K., Zhou, X., Lin, F., Zeng, W., Alterovitz, G.: Deep probabilistic matrix factorization framework for online collaborative filtering. IEEE Access **7**, 56117–56128 (2019). https://doi.org/10.1109/ACCESS.2019.2900698
19. Mishra, R., Rathi, S.: Enhanced dssm (deep semantic structure modelling) technique for job recommendation. J. King Saud Univ. Comput. Inf. Sci. **34**(9), 7790–7802 (Oct 2022)
20. Mu, R.: A survey of recommender systems based on deep learning. IEEE Access **6**, 69009–69022 (2018). https://doi.org/10.1109/ACCESS.2018.2880197
21. Roy, D., Dutta, M.: A systematic review and research perspective on recommender systems. J. Big Data **9**, 1–36 (2022). https://api.semanticscholar.org/CorpusID: 248508374
22. Sedhain, S., Menon, A.K., Sanner, S., Xie, L.: Autorec: autoencoders meet collaborative filtering. In: Proceedings of the 24th International Conference on World Wide Web, pp. 111–112. WWW '15 Companion, Association for Computing Machinery, New York, NY, USA (2015). https://doi.org/10.1145/2740908.2742726
23. Sheikh, A.S., et al.: A deep learning system for predicting size and fit in fashion e-commerce. In: Proceedings of the 13th ACM Conference on Recommender Systems, pp. 110–118. RecSys '19, Association for Computing Machinery, New York, NY, USA (2019). https://doi.org/10.1145/3298689.3347006
24. Shrestha, A., Mahmood, A.: Review of deep learning algorithms and architectures. IEEE Access **7**, 53040–53065 (2019). https://doi.org/10.1109/ACCESS. 2019.2912200
25. Tahmasebi, H., Ravanmehr, R., Mohamadrezaei, R.: Social movie recommender system based on deep autoencoder network using twitter data. Neural Comput. Appl. **33**(5), 1607–1623 (Mar 2021). https://doi.org/10.1007/s00521-020-05085-1
26. Tran, P.H., Nguyen, H.T., Nguyen, N.T.: A hybrid approach for neural collaborative filtering. In: 2020 7th NAFOSTED Conference on Information and Computer Science (NICS), pp. 368–373 (2020)
27. Wang, C.S., Chne, H.C., Chiang, J.H.: Discovering what you cared by intelligent recommender system. In: 2019 International Conference on Technologies and Applications of Artificial Intelligence (TAAI), pp. 1–6 (2019)
28. Wu, C., Wang, J., Liu, J., Liu, W.: Recurrent neural network based recommendation for time heterogeneous feedback. Know.-Based Syst. **109**(C), 90–103 (Oct 2016). https://doi.org/10.1016/j.knosys.2016.06.028

Assessment of LSTM and GRU Models to Predict the Electricity Production from Biogas in a Wastewater Treatment Plant

Pedro Oliveira[1(✉)] , Francisco S. Marcondes[1] , M. Salomé Duarte[2,3] ,
Dalila Durães[1] , Gilberto Martins[2,3] , and Paulo Novais[1]

[1] LASI/ALGORITMI Centre, University of Minho, Braga, Portugal
{pedro.jose.oliveira,francisco.marcondes}@algoritmi.uminho.pt,
{dad,pjon}@di.uminho.pt
[2] CEB - Centre of Biological Engineering, University of Minho, Campus de Gualtar,
4710-057 Braga, Portugal
salomeduarte@ceb.uminho.pt, gilberto.martins@deb.uminho.pt
[3] LABBELS - Associate Laboratory, Braga/Guimarães, Portugal

Abstract. Over the decades, we have faced escalating global energy consumption and its consequential environmental impacts, including climate change and pollution. This study explicitly evaluates the use of Long Short-Term Memory (LSTM) and Gated Recurrent Unit (GRU) models for predicting electricity production from the biogas produced in a Wastewater Treatment Plant (WWTP) in Portugal. WWTPs play an essential role regarding environmental sustainability, namely the potential of biogas in mitigating energy consumption's environmental impact. Also, the work details a comparison between the LSTM and GRU model's performance, applying a grid-search methodology for hyperparameter optimization. The study employs the Root Mean Squared Error (RMSE) as an evaluation metric and uses the sliding window method to transform the problem into a supervised one. After several experiments, the results demonstrate that the LSTM-based model outperforms GRU-based models, achieving an RMSE of 347.9 kWh.

Keywords: Anaerobic Digestion · Deep Learning · Electricity · Time Series · Wastewater Treatment Plants

1 Introduction

In recent decades, energy consumption growth has been an increasingly accentuated worldwide growth [1]. With the increase in energy consumption, it has been possible to observe increasingly severe environmental impacts, such as climate change, air pollution, solid waste disposal, among others [1,2]. Considering the

---

P. Oliveira and F. S. Marcondes—Contributed equally to this paper.

© The Author(s), under exclusive license to Springer Nature Switzerland AG 2024
Á. Rocha et al. (Eds.): WorldCIST 2024, LNNS 986, pp. 64–73, 2024.
https://doi.org/10.1007/978-3-031-60218-4_7

last half century, there has been a continuous increase in electricity consumption, reaching a consumption of around 25,500 terawatt-hours in 2022. During this period, energy consumption practically tripled, with the growth of society and the increase in industrialization boosting the electricity demand [3]. Therefore, nowadays there is an urgent need for alternative and renewable energy sources, such as biogas that can be produced through biological processes such as anaerobic digestion (AD) [4,5]. Among the technologies that wastewater treatment facilities (WWTPs) utilise most frequently to stabilise and reduce the organic content of sludge and boost their energy efficiency is AD. [6].

Through the use of Artificial Intelligence (AI) techniques, these can be widely applied to the various elements of a WWTP [7,8]. One of several application examples includes data-driven energy production forecasting [9]. However, the production of electricity from biogas produced in the anaerobic bioreactors is an area that has been extensively researched [10–12], mainly due to its promising economic performance [13]. Therefore, through AI techniques [14], predicting the energy production of a bioreactor is also under development, without losing sight of explainability [15], to which this article is expected to contribute.

This work aims to evaluate the use of Long Short-Term Memory (LSTM) and Gated Recurrent Unit (GRU) for predicting the electricity produced by the biogas generated in the anaerobic digester of a WWTP. This work considered data collected at a Portuguese municipal wastewater treatment plant. Thus, another objective is to deliver a model capable of predicting energy production as a result of the biogas produced in the biodigester, a continuous stirred tank reactor (CSTR) with a working volume of 2145 m^3 as long as it operates within the same conditions and efficiency rates as described in this paper. The second objective is particularly valuable for allowing the simulation to design and manage WWTPs that employ this bioreactor.

2 Material and Methods

This section describes the materials and methods inherent to the conception of this article, from data collection and exploration to their processing. Furthermore, the Deep Learning (DL) models used in this study are briefly described.

Data Collection. The data used in this study are related to different parameters from a CSTR, operating under anaerobic conditions, and treating municipal sewage sludge. These data were provided by a multi-municipal Portuguese wastewater treatment company, with a collection period between January 2020 and March 2023.

Data Exploration. The dataset analyzed in this study comprises a diverse set of parameters present in the WWTP anaerobic digester reactor. More specifically, there are 14 features in a total of 1186 records. Table 1 summarizes the different features present in the dataset, indicating a short description and measurement units for each feature.

Table 1. Available features in the anaerobic digestion reactor dataset.

#	Feature	Description	Unit of measure
1	date	Timestamp	Date and time
2	Q	Volumetric flow	m^3/day
3	T	Temperature	$^{\circ}C$
4	q_gas_prod	Biogas produced	m^3/day
5	q_gas_cog	Biogas cogenerated	m^3/day
6	p_gas_total	Partial pressure of biogas	bar
7	p_gas_ch4	Partial pressure of methane	bar
8	p_gas_co2	Partial pressure of biogas of carbon dioxide	bar
9	%CO2	percentage of CO2 in the biogas	%
10	%CH4	percentage of CH4 in the biogas	%
11	VS	volatile solids	mg/L
12	X_I	particulate inert matter	mg/L
13	Xc	biodegradable matter	mg/L
14	E	electricity	kWh

The first step in the data exploration stage was to evaluate the existence of missing timestamps and missing values. In this context, it was possible to verify that there were no missing timestamps in the dataset under study. However, it was possible to verify the existence of 7 missing values, whose treatment will be explained in the following subsection. Moreover, we can see that every record in the dataset was associated with a particular day, indicating that each record had a daily frequency.

A statistical analysis was then carried out on the target variable of this study, the electricity feature. Through this analysis, it was possible to find out that this feature's average value was 1575.67 kWh, with a standard deviation of 783.37 kWh. The minimum and maximum values were 0 kWh and 4975 kWh, respectively. Furthermore, we analysed two other metrics, the skewness and kurtosis values. Concerning skewness, this variable presented a value of -40.030, representing a negatively asymmetric distribution. On the other hand, the kurtosis value was 0.166, thus showing a leptokurtic distribution.

Finally, we checked that the features in this dataset followed a normal distribution. The Kolmogorov-Smirnovc test was used for this, with a $p < 0.05$. The obtained results made it possible to conclude that all features did not present a normal distribution. This evaluation impacted the chosen methodology, shown in the following subsection, for feature correlation analysis.

Data Preparation. The first step in the data processing stage was to fill in the missing values detected in the data analysis phase, namely concerning the target of this study, electricity. Considering that the number of missing values verified was widely spaced throughout the dataset, the linear interpolation method was applied to fill them.

Then, it was verified that there were 78 lines in the dataset with both *q_gas_cog* and *electricity* equal to zero. The most problematic period was between the end of 11/2020 and the beginning of 01/2021, when the values were almost continuously zero. Since these values were considered necessary to keep all data meaningful, different steps were followed for this treatment. The first step was to deal with the zeros in the feature *q_gas_cog*, replacing them, randomly, with values above or below the average value of this feature, taking into account the value of its standard deviation. Next, considering that energy use is directly related to biogas composition, whose energy potential generally varies between 6 and 6.4 kWh/m^3. Typically, the electrical efficiency of a cogeneration installation varies between 25% and 45%, while the thermal efficiency varies between 45% and 60% concerning the amount of methane present in the biogas [16]. In this sense, the average efficiency of the present dataset data was calculated considering the Eq. 1:

$$\frac{1}{n} \sum_{i=1}^{n} \frac{\text{electricity}}{\text{cogenerator\_gas}} \tag{1}$$

Through Eq. 1, we concluded that average energy efficiency assumed a value of 1.65 kWh/m3. With this calculated value, it was then possible to deal with the zeros present in the *electricity* feature. To do this, in each observation where the electricity value was 0, the average energy efficiency value calculated was multiplied by the cogenerator gas value.

After carrying out the replacement processing of the different zeros verified, Fig. 1 was developed, which shows the data distribution regarding the cogenerator gas in the different timesteps, considering the electricity value. By analysing the figure, we can see that the highest concentration of data regarding gas was between the fourth and second quartile, considering the electricity values. Furthermore, it was also possible to observe that, globally, the increase in the value of electricity was associated with the increase in the value of biogas from the cogenerator.

Next, it was necessary to analyse the correlation between the different features present in the dataset with the target of this study, electricity from the biogas produced in the anaerobic reactor. Considering that the different features in the dataset did not present a Gaussian distribution, as analysed in the data exploration section, the Spearman correlation test coefficient was used. By analysing the results obtained using this coefficient, it was possible to verify that only the features *q_gas_cog*, *T* and *q_gas_prod* present significant correlations with electricity. In this sense, all remaining features were removed, leaving the final dataset, after all the treatments, with four features, the three presented with correlation with the target and the respective target.

Finally, the last step in the data processing stage was transforming the problem into a supervised problem, considering the sliding window method. This process is essential to create input/label pairs, that is, the X and the y, depending on the number of time steps that make up a sequence. Furthermore, for the LSTM-based candidate models, the data were normalised between −1 and 1, while for GRU-based candidate models, they were normalised between 0 and 1.

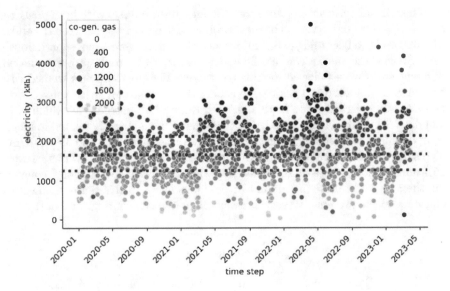

Fig. 1. Electricity produced regarding the co-generated gas. The blue lines mark, respectively the $4^{th}, 3^{rd}$ and 2nd quartiles.

Deep Learning Models. It is beyond the scope of this paper to detail the Long Short-Term Memory (LSTM) and Gated Recurrent Unit (GRU) approaches. In short, both are suitable for approaching time series problems because they solve the vanishing gradient problem by preserving long-range dependencies. See Fig. 2 for a reference. LSTM has a more complex structure and performs better than GRU on long term dependencies. In contrast, GRU is less expensive than LSTM and is a suitable solution when it converges.

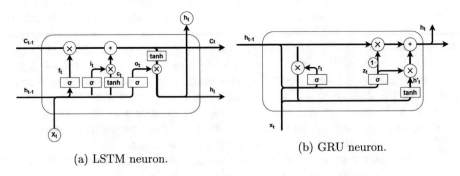

(a) LSTM neuron.

(b) GRU neuron.

Fig. 2. LSTM and GRU neuron architecture [17]. The '×' symbol is a pointwise multiplication and the '+' symbol a pointwise addition.

3 Experiments

In order to find which of the two models under study presented better performance and the associated set of hyperparameters, the grid-search methodology was applied to a vast set of hyperparameters in each DL model. Table 2 presents the different hyperparameter values used. Furthermore, the evaluation metrics used to verify the performance of the different candidate models are presented.

Table 2. Hyperparameters' searching space.

Parameters	Values
Layers	[3,4,5]
Neurons	[32,64,128]
Dropout Rate	[0.0,0.5]
Activation Function	[ReLU,tanh]
Timesteps	[7,14,21]
Batch Size	[5,10,20]

Each candidate model was evaluated by the Root Mean Squared Error (RMSE) given by the expression:

$$\text{RMSE} = \sqrt{\frac{\sum_{i=1}^{n}(yi - yi_{pred})^2}{n}} \tag{2}$$

To put it briefly, it evaluates the correctness of the model based on the average variance of the regression's residuals.

One of the aspects taken into account in the development of the experiments was to guarantee the reproducibility of the obtained results in this study. In this context, all experiments developed used the same seed value, specifically 91195003.

Furthermore, the learning curves for each DL model were analysed to prevent overfitting or overfitting situations. Analysing these curves made it possible to verify and choose the number of epochs to use in each model. Considering the analysis carried out, in the case of GRU-based models, the value was 70 epochs, while in LSTM-based models, the number of epochs selected was 65. Moreover, a cross-validation technique was adopted so the candidate models do not generalise. As in this specific study, we are dealing with models considering time series problems, it was necessary to use a cross-validator considering this problem. Therefore, the cross-validator used in this study was *TimeSeriesSplit*.

4 Results and Discussion

After carrying out all the experiments previously described and conceiving all the candidate models, it was necessary to analyze the obtained results. In this

Table 3. Top-five candidate models results, for LSTMs and GRUs.

Timesteps	Batch	Layers	Neurons	Dropout	Activation	Epochs	RMSE	Time(s)
			LSTMs candidate models					
7	5	5	128	0.0	relu	65	347.9	309.7
7	5	5	64	0.5	relu	65	362.8	307.9
7	5	3	64	0.5	relu	65	388.9	227.9
7	20	5	64	0.5	relu	65	391.1	141.9
7	10	5	32	0.0	relu	65	395.7	183.5
			GRUs candidate models					
7	5	4	32	0.5	relu	70	394.8	320.2
7	5	5	32	0.5	tanh	70	394.9	266.9
7	5	3	32	0.5	relu	70	399.5	236.4
7	10	5	32	0.0	relu	70	401.1	170.4
7	5	3	64	0.5	relu	70	401.4	238.1

sense, Table 3 summarizes the results of the best 5 candidate models for each approach, LSTMs and GRUs.

With an analysis of the previous table regarding the results obtained, it is possible to evidence that the best candidate model is the LSTM-based model, with an RMSE value of 347.9 kWh, which represents an error percentage of 7% considering the data used in this study. Regarding the LSTM-based candidate models, it is possible to see in the five identified in the table that the best candidate is the one with the highest number of neurons, with 128, compared to the remaining LSTM-based, which mainly uses 64 neurons. Furthermore, it is possible to verify that the 5 LSMT-based candidate models require a homogeneous number of timesteps and activation functions, in this case, 7 and relu, respectively, to predict the next two days regarding energy consumption in the anaerobic reactor digester. Regarding the number of intermediate layers, most candidate models required 5 layers, except one candidate model.

Regarding the GRU-based candidate models, we can see that the best candidate model has an RMSE value of 394.8 kWh, approximately 7.9% error percentage, with 32 neurons and 4 intermediate layers. When comparing the best GRU-based candidate model with the others, we verify that there is homogeneity in most of its hyperparameters, as in the case of the *drop out* layer rate (0.5) and number of timesteps (7). In the remaining hyperparameters, despite not being homogeneous like the others, their heterogeneity is relatively low, with only a few modifications.

In a broader comparison between the LSTM-based and GRU-based candidate models, we can conclude that all candidate models need 7 timesteps, i.e., the last 7 d, to predict the next two days concerning electricity in the cogenerator. Furthermore, r Regarding RMSE, although the best LSTM-based candidate model, that is the best of all, there is a more pronounced decrease compared to other LSTM-based candidates in RMSE. On the other hand, in the 5 best GRU-based candidates, the error decrease between candidate models is not as pronounced. Regarding the training time between the best candidate models of both approaches, the difference between them is around 10 s, with no significant pattern between them or among the other candidate models that can be highlighted in terms of model training time.

5   Conclusions

WWTPs are essential in mitigating pollution by treating and purifying water before it's released into the environment. However, these facilities often consume significant energy in their operations. The biogas, and ultimately energy, produced in the anaerobic digestion of sewage sludge in WWTP may not only help to reduce the environmental impact of the energetic needs of these infrastructures but also contribute to the broader goal of achieving energetic sustainability.

Throughout this study, several experiments were carried out, with the conception of different candidate models, both LSTM-based and GRU-based, to predict energy production from biogas produced through anaerobic digester in a WWTP. Taking into account the data processing carried out through the calculation of energy efficiency to fill the electricity values to zero, the experiments aimed to find which of the two models under study presented the best performance in predicting electricity generated by an anaerobic digestor process and which is the best combination of hyperparameters in each of them. By analysing the obtained results, it was possible to conclude that the best candidate model, in terms of RMSE, was LSTM-based with an error value of 347.9 kWh.

Given that anaerobic bioreactors are fairly stable systems, it is possible to use this model to predict electricity production, given the same constraints. This result is of interest to the general public, as this type of solution is increasingly being used.

Regarding future work, we focus on implementing more Machine Learning models, in this case, Multi-Layer Perceptron and Convolutional Neural Networks. Furthermore, the use of hybrid models using Transformers will also be taken into account in order to understand whether they will obtain better performances than the models used in this study. Another point to be considered will be adopting new approaches to treating the zero values presented in the target to be predicted in this study, realizing the impact that different approaches to treating these values can have on the performance of the conceives models.

Acknowledgements. This work is financed by National Funds through the Portuguese funding agency, FCT - Fundação para a Ciência e a Tecnologia within project 2022.06822.PTDC. The work of Pedro Oliveira was supported by the doctoral Grant PRT/BD/154311/2022 financed by the Portuguese Foundation for Science and Technology (FCT), and with funds from European Union, under MIT Portugal Program.

References

1. Ahmad, T., Zhang, D.: A critical review of comparative global historical energy consumption and future demand: The story told so far. Energy Rep. **6**, 1973–1991 (2020). https://doi.org/10.1016/j.egyr.2020.07.020

2. Vertakova, Y.V., Plotnikov, V.A.: The integrated approach to sustainable development: the case of energy efficiency and solid waste management. Int. J. Energy Econ. Policy **9**(4), 194–201 (2019). https://doi.org/10.32479/ijeep.8009

3. Scheffran, J., Felkers, M., Froese, R.: Economic growth and the global energy demand. Green Energy Sustain.: Strat. Global Indust. 1–44 (2020). https://doi.org/10.1002/9781119152057.ch1

4. Rosén, C., Vrecko, D., Gernaey, K.V., Pons, M.N., Jeppsson, U.: Implementing adm1 for plant-wide benchmark simulations in matlab/simulink. Water Sci. Technol. **54**(4), 11–19 (2006)

5. Obileke, K.C., Nwokolo, N., Makaka, G., Mukumba, P., Onyeaka, H.: Anaerobic digestion: technology for biogas production as a source of renewable energy-a review. Energy Environ. **32**(2), 191–225 (2021). https://doi.org/10.1177/0958305X20923117

6. Cardoso, B.J., Rodrigues, E., Gaspar, A.R., Gomes, A.: Energy performance factors in wastewater treatment plants: a review. J. Cleaner Prod. **322** 129107 (2021). https://doi.org/10.1016/j.jclepro.2021.129107

7. Kamali, M., Appels, L., Yu, X., Aminabhavi, T.M., Dewil, R.: Artificial intelligence as a sustainable tool in wastewater treatment using membrane bioreactors. Chem. Eng. J. **417**, 128070 (2021). https://doi.org/10.1016/j.cej.2020.128070

8. Singh, N.K., et al.: Artificial intelligence and machine learning-based monitoring and design of biological wastewater treatment systems. Bioresource technol. **369** 128486 (2022). https://doi.org/10.1016/j.biortech.2022.128486.

9. Harrou, F., Cheng, T., Sun, Y., Leiknes, T.O., Ghaffour, N.: A data-driven soft sensor to forecast energy consumption in wastewater treatment plants: a case study. IEEE Sens. J. **21**(4), 4908–4917 (2020). https://doi.org/10.1109/JSEN.2020.3030584

10. Pierangeli, G.M.F., Ragio, R.A., Benassi, R.F., Gregoracci, G.B., Subtil, E.L.: Pollutant removal, electricity generation and microbial community in an electrochemical membrane bioreactor during co-treatment of sewage and landfill leachate. J. Environ. Chem. Eng. **9**(5), 106205 (2021). https://doi.org/10.1016/j.jece.2021.106205.

11. Li, T., Cai, Y., Yang, X.-L., Yan, W., Yang, Y.-L., Song, H.-L.: Microbial fuel cell-membrane bioreactor integrated system for wastewater treatment and bioelectricity production: overview. J. Environ. Eng. **146**(1), 04019092 (2020). https://doi.org/10.1061/(ASCE)EE.1943-7870.000160

12. Ibrahim, R.S.B., Zainon Noor, Z., Baharuddin, N.H., Ahmad Mutamim, N.S., Yuniarto, A.: Microbial fuel cell membrane bioreactor in wastewater treatment, electricity generation and fouling mitigation. Chem. Eng. Technol. **43**(10), 1908–1921 (2020). https://doi.org/10.1002/ceat.202000067

13. Vinardell, S., Dosta, J., Mata-Alvarez, J., Astals, S.: Unravelling the economics behind mainstream anaerobic membrane bioreactor application under different plant layouts. Biores. Technol. **319**, 124170 (2021). https://doi.org/10.1016/j.biortech.2020.124170

14. Pereira, J., Oliveira, P., Duarte, M.S., Martins, G., Novais, P.: Using deep learning models to predict the electrical conductivity of the influent in a wastewater

treatment plant. In: International Conference on Intelligent Data Engineering and Automated Learning, pp. 130–141. Springer (2023). https://doi.org/10.1007/978-3-031-48232-8_13

15. Marcondes, F.S., Durães, D., Santos, F., Almeida, J.J., Novais, P.: Neural network explainable AI based on paraconsistent analysis: an extension. Electronics **10**(21), 2660 (2021)

16. Moses Jeremiah Barasa Kabeyi and Oludolapo Akanni Olanrewaju: Technologies for biogas to electricity conversion. Energy Rep. **8**, 774–786 (2022). https://doi.org/10.1016/j.egyr.2022.11.007

17. Oliveira, P., Fernandes, B., Analide, C., Novais, P.: Forecasting energy consumption of wastewater treatment plants with a transfer learning approach for sustainable cities. Electronics **10**(10), (2021). https://doi.org/10.3390/electronics10101149

Fusing Temporal and Contextual Features for Enhanced Traffic Volume Prediction

Sara Balderas-Díaz$^{(\boxtimes)}$, Gabriel Guerrero-Contreras, Andrés Muñoz, and Juan Boubeta-Puig

Department of Computer Science and Engineering, University of Cadiz, Av. Universidad de Cádiz, 10, 11519 Puerto Real, Cádiz, Spain
{sara.balderas,gabriel.guerrero,andres.munoz,juan.boubeta}@uca.es

Abstract. Time-series analysis plays a crucial role in extracting meaningful patterns from sequential data, serving to gain insights into temporal trends. Specifically, the burgeoning urbanization trend accentuates the urgency of traffic forecasting in modern cities due to its socio-economic and environmental impact. While Deep Learning techniques, notably Long Short-Term Memory (LSTM) networks, have shown promise in predicting traffic, they often overlook contextual factors like weather and holidays. To address this challenge, this paper proposes a hybrid model combining LSTM with categorical and continuous data to forecast traffic volume on the Interstate 94 American highway. Incorporating weather, temporal patterns, and holidays, this study explores the performance of a model that includes contextual factors against standalone LSTM. The results show superior predictive accuracy for the proposed hybrid model. SHapley Additive exPlanations (SHAP) analysis reveals the influence of diverse features, emphasizing the significance of contextual attributes in enhancing traffic prediction.

Keywords: time-series analysis · traffic forecasting · hybrid neural network · LSTM · contextual factors

1 Introduction

Time-series analysis encompasses various tasks with the goal of deriving valuable insights from sequentially ordered data. This extracted knowledge serves the dual purpose of retrospectively diagnosing past behaviors and predicting proactively future trends [1]. The application of time-series forecasting extends to diverse real-world domains, including stock prediction, finance, agriculture, healthcare, transportation, climate, and air quality [2,3].

In today's modern cities, traffic forecasting has become a pressing issue because traffic congestion leads to lost time, increased pollution, and economic

G. Guerrero-Contreras, A. Muñoz, J. Boubeta-Puig—Contributing authors.

inefficiencies [4]. These problems may be aggravated, considering that 55% of the population currently lives in urban areas and that in the next four decades, a global population increase of 2.5 billion is expected [5]. In this regard, the adoption of Machine Learning (ML) and Deep Learning (DL) techniques to predict traffic issues is on the rise. Indeed, by leveraging the capabilities of these artificial intelligence models, it is possible to predict traffic effectively [6,7].

Nevertheless, traditional traffic prediction models often ignore the influence of contextual factors such as holidays or weather conditions (i.e., snow, rain, cloud cover, and temperature, among others). To address this limitation, our study proposes a hybrid model that combines both categorical (i.e., holiday periods, weekday, snow, rain, etc.) and continuous data (i.e., temperature, historic traffic volume, etc.) in a Dense Neural Network (DNN) with an input from a Long Short-Term Memory (LSTM) network with the objective of predicting future traffic volume. This research focuses on the analysis of contextual attributes in predicting traffic volume in the time-series data of the Interstate 94 highway in United States.

The paper is organized as follows: Sect. 2 reviews recent time-series proposals for predicting traffic volume. Section 3 details the dataset, data processing, and architecture of the proposed neural networks. Section 4 discusses the outcomes, and Sect. 5 summarizes the study's conclusions.

2 Related Work

Traffic management in urban areas remains a pressing concern, prompting innovative approaches leveraging technology. In [8] the limitations of traditional models are analyzed and a hybrid method is proposed, ST-LSTM, integrating Savitzky-Golay (SG) filtering, Temporal Convolutional Networks (TCN), and LSTM. This approach effectively captures short and long-term dependencies, outperforming existing models, and offers valuable insights into handling modern network traffic characteristics.

Emphasizing road safety for Vulnerable Road Users (VRUs), in [9], AutoRegressive Integrated Moving Average (ARIMA) is compared with LSTM algorithms for traffic flow prediction. It demonstrates LSTM superiority, especially in data-rich environments, reaffirming its potential for accurate predictions in urban settings.

In [10], a genetic algorithm-based optimization approach for traffic signal controllers in urban areas is presented. It poses a solution for enhancing traffic flow efficiency, reducing congestion, and improving environmental conditions.

Moreover, in predictive business process monitoring, the study in [11] emphasized the importance of context-aware analysis. Their novel approach incorporates context attributes to significantly enhance completion time prediction models, showcasing substantial improvements in predictive accuracy compared to existing techniques.

The examined studies emphasize the rising importance of sophisticated algorithms such as LSTM and hybrids across various fields like urban planning, traffic management, safety, emergencies, and business optimization. They emphasize the need for context-awareness in predictive analytics but overlook the incorporation of specific contextual elements like holidays and weather conditions, a focal point in our proposed approach.

3 Methodology

This section details the methodology employed in this work, encompassing dataset description, data preprocessing, modeling techniques, and evaluation metrics. Figure 1 illustrates the architecture and the process implemented for the traffic volume prediction system. The model comprises an LSTM layer capturing temporal dependencies, and dense layers fusing LSTM output with encoded categorical information.

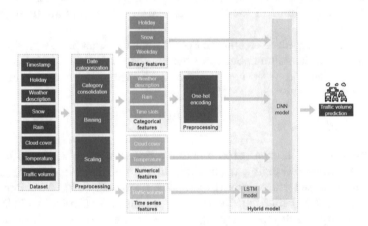

Fig. 1. Architectural representation of the hybrid model integrating LSTM with categorical features for traffic volume prediction.

3.1 Dataset

For the study, the Metro Interstate Traffic Volume dataset[1] has been selected. This dataset captures the traffic volume on the westbound lanes of Interstate 94 in the United States. It gathers this information through an Automated Traffic Recorder (ATR) station, specifically, the station 301, overseen by the Minnesota Department of Transportation (MnDOT). Positioned between Minneapolis and St. Paul, this station provides detailed hourly records. Moreover, the dataset includes factors that could potentially impact traffic volume, such as weather conditions and holiday occurrences. Each entry contains the following attributes:

- `holiday`: A categorical feature denoting US National holidays in addition to a regional holiday, specifically the Minnesota State Fair.

---

[1] https://archive.ics.uci.edu/dataset/492/metro+interstate+traffic+volume.

- `temp`: A numeric attribute representing the average temperature in Kelvin for each recorded hour.
- `rain_1h`: A numeric attribute indicating the quantity of rainfall, measured in millimeters, within the hour.
- `snow_1h`: A numeric attribute specifying the amount of snowfall, measured in millimeters, observed within the hour.
- `clouds_all`: A numeric attribute expressing the percentage of cloud cover during the observed hour.
- `weather_main`: A categorical attribute providing a textual summary of the prevailing weather conditions.
- `weather_description`: A categorical attribute offering a detailed textual description of the observed weather conditions.
- `date_time`: A DateTime attribute representing the timestamp of data collection in local CST time.
- `traffic_volume`: A numeric attribute indicating the hourly I-94 ATR 301 reported westbound traffic volume.

The dataset comprises a total of 48,204 instances recorded within the period spanning 2012 to 2018. Despite the absence of null values in the dataset, entries presenting zeros across all attributes were deemed erroneous and consequently eliminated from the dataset. This action resulted in a final dataset size of 48,193 records.

3.2 Data Preprocessing

In this section, a series of preprocessing steps were implemented to optimize the dataset for subsequent analysis and modeling.

The dataset encompasses 11 unique holidays, including *Labor Day, Thanksgiving Day*, and *Christmas Day*. Among all recorded days, merely 50 of them acknowledge any of these holidays. For streamlined analysis and improved efficiency, a category consolidation process was conducted. As a result, days are marked as *True* if any holiday is observed and *False* if not, irrespective of the specific holiday type.

Likewise, within the `weather_description` attribute, there are 38 distinct categories. Some of these demonstrate infrequent occurrences, like *light rain and snow* and *shower drizzle*, each with 6 counts, *squalls* with 4 counts, *sleet* with 3 counts, *freezing rain* and *thunderstorm with drizzle*, each with 2 counts, and *shower snow* with 1 count. Consequently, these categories have been reorganized into broader classifications as shown in Table 1.

Concerning the `snow_1h` attribute, only 63 instances reveal snowfall greater than zero. Consequently, a binning method was applied, labeling it as *snow* during hours with recorded snowfall and *no_snow* otherwise. Similarly, for `rain_1h`, a binning process was executed. Entries with a value of zero were classified as *no_rain*. For records with `rain_1h` greater than zero, quartiles were used to categorize values: those up to the 25th percentile were classified as *very_light_rain*,

Table 1. Categories consolidation in `weather_description` attribute.

Category group	Category	Category count	Group count
clear	sky is clear	13,381	13,381
scattered	few clouds	1,956	15,164
	broken clouds	4,666	
	overcast clouds	5,081	
	scattered clouds	3,461	
light rain	drizzle	651	5,135
	light intensity drizzle	1,100	
	light rain	3,372	
	light rain and snow	6	
	shower drizzle	6	
moderate rain	moderate rain	1,664	1,664
heavy rain	heavy intensity drizzle	64	548
	heavy intensity rain	467	
	very heavy rain	17	
snow	snow	293	2,867
	light snow	1,946	
	shower snow	1	
	heavy snow	616	
	light shower snow	11	
special	fog	912	8,247
	freezing rain	2	
	haze	1,360	
	mist	5,950	
	smoke	20	
	sleet	3	
storm	thunderstorm	125	1,034
	proximity thunderstorm	673	
	proximity thunderstorm with drizzle	13	
	proximity thunderstorm with rain	52	
	thunderstorm with drizzle	2	
	thunderstorm with heavy rain	63	
	thunderstorm with light drizzle	15	
	thunderstorm with light rain	54	
	thunderstorm with rain	37	
special precipitation	light intensity shower rain	13	153
	proximity shower rain	136	
	squalls	4	

values up to the 50th percentile as *light_rain*, up to the 75th percentile as *moderate_rain*, and values beyond the 75th percentile as *heavy_rain*.

Moreover, two new features were introduced to capture temporal patterns. Thus, is_weekday distinguishes weekdays (Monday to Friday) from weekends (Saturday or Sunday) based on the date_time. Additionally, time_slot categorizes hours of the day into specific intervals: *night-early morning* (21:00–6:00), *morning peak hours* (6:00–9:00), *morning-noon* (9:00–12:00), *noon-afternoon peak hours* (12:00–15:00), *late afternoon hours* (15:00–18:00) and *evening-night* (18:00–21:00). These preprocessing steps aim to accurately capture temporal patterns, crucial for modeling traffic behaviors that vary throughout the day and week.

Regarding numerical attributes (traffic_volume, clouds_all, and temp), which showed no outliers (see Fig. 2), they were scaled between 0 and 1 using a min-max approach. This scaling guarantees a consistent scale across all numerical variables, aiding interpretation and model training. Min-max has been applied after splitting the data into training and testing.

3.3   LSTM for Traffic Data Prediction

LSTM networks are a type of Recurrent Neural Network (RNN) designed to handle long-range dependencies and sequential data. They are particularly effective in modeling time series data due to their ability to retain and selectively forget information over varying time intervals.

The core idea behind LSTMs lies in their gated architecture, which consists of memory cells and gates to regulate the flow of information. These gates include an input gate (i_t), a forget gate (f_t), an output gate (o_t), and the cell state (C_t).

The computations within an LSTM cell at time step t are defined by Eq. 1.

$$
\begin{aligned}
i_t &= \sigma(W_i \cdot [h_{t-1}, x_t] + b_i) \\
f_t &= \sigma(W_f \cdot [h_{t-1}, x_t] + b_f) \\
o_t &= \sigma(W_o \cdot [h_{t-1}, x_t] + b_o) \\
\tilde{C}_t &= \tanh(W_C \cdot [h_{t-1}, x_t] + b_C) \\
C_t &= f_t \odot C_{t-1} + i_t \odot \tilde{C}_t \\
h_t &= o_t \odot \tanh(C_t)
\end{aligned}
\tag{1}
$$

Here, x_t denotes the input at time t, h_t represents the hidden state, W_i, W_f, W_o, W_C are weight matrices, b_i, b_f, b_o, b_C are bias vectors, and σ denotes the sigmoid function.

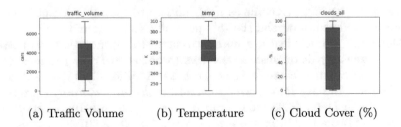

|(a) Traffic Volume | (b) Temperature | (c) Cloud Cover (%) |

Fig. 2. Distribution of `traffic_volume`, `clouds_all`, and `temp` recorded in dataset.

3.4   LSTM and Categorical Features Integration in a Hybrid Neural Network

This model combines the strengths of LSTM networks in handling temporal dependencies with the incorporation of categorical information (Fig. 1). The model architecture involves the use of an LSTM layer to capture temporal patterns from the sequential data. Simultaneously, categorical attributes such as `holiday`, `weather_main`, and `weather_description` are processed separately.

The LSTM layer extracts temporal patterns from the sequential data and produces an output representation capturing the learned dependencies over time. This output is then concatenated with the encoded categorical features before feeding into a neural network. This network is designed to fuse the LSTM output with the categorical features, enabling the model to learn complex relationships between the temporal and categorical data.

One-hot encoding converts categorical variables into a binary format, creating a binary vector for each category within the variable. The process involves creating new columns, where each column corresponds to a unique category within the categorical variable. For instance, considering the `weather_main` attribute with categories like *clouds, clear, rain,* and others, one-hot encoding would transform this into separate binary columns. Each row represents an entry in the dataset, and the value '1' in a specific category column indicates the presence of that category in the original categorical variable for that entry.

3.5   Evaluation Metrics

In this research, two common evaluation metrics are used to assess the performance of the models, Root Mean Squared Error (RMSE) (Eq. 2) and Mean Absolute Error (MAE) (Eq. 3).

$$\text{RMSE} = \sqrt{\frac{1}{n}\sum_{i=1}^{n}(y_i - \hat{y}_i)^2} \quad (2) \qquad \text{MAE} = \frac{1}{n}\sum_{i=1}^{n}|y_i - \hat{y}_i| \quad (3)$$

where n is the number of samples, y_i represents the actual values, and $\hat{y}_i$ represents the predicted values.

4   Experimental Results

In this section, the experimental results of the LSTM and hybrid models are presented, showing their distinct architectures and performance metrics.

4.1   Training

In our study, the LSTM and hybrid models showcase distinct architectures. The LSTM model integrates a single LSTM layer featuring 130 hidden units. Its input remains univariate with a dimensionality of 1. The *tanh* activation function is used for recurrent operations, while the *sigmoid* activation function manages input and forget gates. A look-back window of 24 h is used to predict the subsequent time slot. Conversely, the hybrid model comprises three dense layers. These layers adopt configurations of 256 and 512 units successively, employing ReLU activation. The final layer, responsible for predicting `traffic_volume`, hosts a single unit. For both architectures, the Adam optimizer is used with a fixed learning rate of 0.001. After conducting several experiments, these parameters have proven to be the most suitable.

The dataset is partitioned into training and testing sets following a 70–30 split. To ensure robustness and validate the generalization ability of models, a 5-fold cross-validation approach is also employed during the training phase, with a 64-batch size.

4.2   Performance Comparison for Traffic Volume Prediction

In terms of RMSE, the LSTM model achieved a value of 407.03, indicating the average magnitude of the errors between predicted and actual values. Conversely, the hybrid model exhibited a lower RMSE of 378.61, representing a superior performance in minimizing prediction errors. Similarly, when considering MAE,

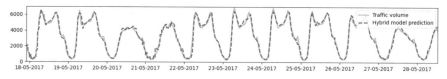

(a) Forecast traffic volume data by the hybrid model and the actual observed values.

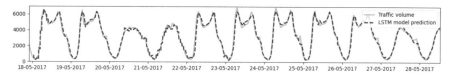

(b) Forecast traffic volume data by LSTM model and the actual observed values.

Fig. 3. Prediction comparison for models developed on a 14-day sample of the test set.

the LSTM model recorded an average error of 285.45, while the hybrid model notably outperformed it with a lower MAE of 256.51. These results suggest that the hybrid model, integrating both LSTM and additional categorical features, obtains improved accuracy in predicting traffic volume compared to the standalone LSTM model. Figure 3 shows a visual representation of the forecast traffic volume data by the hybrid and the LSTM models, alongside the actual observed values. The comparison is based on a 14-day sample of the test set.

4.3 Feature Influence Analysis

The SHapley Additive exPlanations (SHAP) analysis method has been applied to uncover the influence of each feature in the hybrid model [12]. It operates based on the game theory, assigning a SHAP value to each feature, indicating its impact on the difference between the prediction of the model and a baseline prediction.

In the hybrid model, the feature ranking (see Fig. 4), arranged by descending order of impact in the predictive capacity of the model is: is_weekday, time_slot, temp, weather_main, clouds_all, weather_description, holiday, rain_1h, and snow_1h.

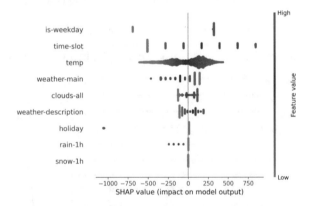

Fig. 4. Influence of additional features on hybrid model predictions.

Weekdays (is_weekday) notably drive increased traffic volumes in contrast to weekends. Moreover, daytime slots (time_slot) display heightened traffic, whereas nighttime periods register lower levels. The third most influential feature, temp, showcases a correlation between increased temperature and higher traffic, likely indicating heightened daytime activities as the feature time_slot indicates. Conversely, lower temperatures, associated with adverse weather or nighttime, correspond to decreased traffic. Similarly, weather_main indicates higher traffic during normal conditions like *clouds, clear, rain, drizzle,* or *mist,* while extreme weather conditions like *haze, fog, thunderstorm, snow, squall,* or *smoke* result in reduced traffic. `clouds_all` and `weather_description` display similar trends, as they are directly linked to `temp` and `weather_main` features.

Moreover, the `holiday` feature, despite its reduced occurrence, significantly contributes to predict reduced traffic, as well as, the `rain_1h` feature during heavy rainfall events. Finally, `snow_1h` exhibits minimal influence on the model, likely due to the limited instances (63) recording this value as *True*.

5   Conclusion and Future Work

This study introduced a hybrid neural network, merging LSTM with categorical features for traffic volume prediction. Using a rich dataset encompassing traffic, weather, and time attributes, the model effectively grasped intricate temporal-contextual relationships. It notably outperformed standalone LSTM, exhibiting improved predictive accuracy with lower RMSE and MAE values. This approach may aid better traffic resource planning amid significant traffic fluctuations in specific areas.

Future research directions may explore the incorporation of additional contextual features, such as road maintenance schedules or special events, to further enhance the model's predictive capabilities. Moreover, investigating advanced deep learning architectures beyond LSTM, such as attention mechanisms, could offer deeper insights into capturing long-term dependencies in traffic data. Additionally, the proposal will be enriched with Complex Event Processing (CEP) techniques to automatically detect new meaningful causality, membership and temporal relationships between datasets and real-time data from smart cities such as traffic, weather, and air quality. Thanks to the use of CEP, the integration of historical time series and real-time streaming data into an all-in-one solution will become a reality.

Acknowledgments. Financial support for this research has been provided under AwESOMe Project PID2021-122215NB-C33 and ALLEGRO Project PID2020-112827GB-I00, both funded by MCIN/ AEI /10.13039/501100011033/ and by ERDF A way to do Europe.

References

1. Choi, K., Yi, J., Park, C., Yoon, S.: Deep learning for anomaly detection in time-series data: review, analysis, and guidelines. IEEE Access **9**, 120043–120065 (2021)
2. Zhang, H., Sun, B., Peng, W.: A novel hybrid deep fuzzy model based on gradient descent algorithm with application to time series forecasting. Expert Syst. Appl. **238**, 121988 (2024)
3. Méndez, M., Merayo, M.G., Núñez, M.: Machine learning algorithms to forecast air quality: a survey. Artif. Intell. Rev. **56**(9), 10031–10066 (2023)
4. Boukerche, A., Tao, Y., Sun, P.: Artificial intelligence-based vehicular traffic flow prediction methods for supporting intelligent transportation systems. Comput. Netw. **182**, 107484 (2020)
5. Ritchie, H., Roser, M.: Urbanization. Our World Data (2018)

6. Medina-Salgado, B., Sanchez-DelaCruz, E., Pozos-Parra, P., Sierra, J.E.: Urban traffic flow prediction techniques: a review. Sustain. Comput. Inform. Syst. **35**, 100739 (2022)

7. Muñoz, A., Martínez-España, R., Guerrero-Contreras, G., Balderas-Díaz, S., Bueno-Crespo, A.: A real-time traffic alert system based on image recognition: a case of study in spain. In: 2023 19th International Conference on Intelligent Environments (IE), pp. 1–7 (2023). IEEE

8. Bi, J., Zhang, X., Yuan, H., Zhang, J., Zhou, M.: A hybrid prediction method for realistic network traffic with temporal convolutional network and LSTM. IEEE Trans. Autom. Sci. Eng. **19**(3), 1869–1879 (2021)

9. Fernandes, B., et al.: Traffic flow forecasting on data-scarce environments using ARIMA and LSTM networks. In: Rocha, Á., Adeli, H., Reis, L.P., Costanzo, S. (eds.) WorldCIST'19 2019. AISC, vol. 930, pp. 273–282. Springer, Cham (2019). https://doi.org/10.1007/978-3-030-16181-1_26

10. Balderas-Díaz, S., Muñoz, A., Guerrero-Contreras, G.: Optimization of traffic light controllers using genetic algorithms: a case study in the city of cádiz. In: International Symposium on Ambient Intelligence, pp. 103–112 (2023). Springer https://doi.org/10.1007/978-3-031-43461-7_11

11. Alves, R.M., Barbieri, L., Stroeh, K., Peres, S.M., Madeira, E.R.M.: Context-aware completion time prediction for business process monitoring. In: World Conference on Information Systems and Technologies, pp. 355–365 (2022). Springer https://doi.org/10.1007/978-3-031-04819-7_35

12. Antwarg, L., Miller, R.M., Shapira, B., Rokach, L.: Explaining anomalies detected by autoencoders using shapley additive explanations. Expert Syst. Appl. **186**, 115736 (2021)

Target-vs-One and Target-vs-All Classification of Epilepsy Using Deep Learning Technique

Adnan Amin[1,3], Feras Al-Obeidat[2], Nasir Ahmed Algeelani[3], Ahmed Shudaiber[2], and Fernando Moreira[4(✉)]

[1] Center for Excellence in Information Technology, Institute of Management Sciences, Peshawar, Pakistan
adnan.amin@imsciences.edu.pk
[2] College of Technological Innovation, Zayed University, Abu Dhabi, UAE
{feras.al-obeidat,ahmed.shuhaiber}@zu.ac.ae
[3] Faculty of Computer Science and Information Technology, Al-Madinah International University (MEDIU), Kuala Lumpur, Malaysia
nasir.ahmed@mediu.edu.my
[4] REMIT, Universidade Portucalense, Porto and IEETA, Universidade de Aveiro, Aveiro, Portugal
fmoreira@upt.pt

Abstract. With the pervasive generation of medical data, there is a need for the worldwide medical and health care sector to find appropriate computational intelligence techniques for various medical conditions such as epilepsy seizures (ES). ES is a brain disorder that affects people of all ages, is a chronic, non-communicable disease, and can occur for no apparent reason owing to a genetic defect at any time. The unpredictable nature of ES poses a significant threat to human life where we have a target variable with five labels of seizure, namely pre-seizure, seizure and seizure-free, are classified using real clinical data. In order to accurately classify seizure activity (e.g., the target label) without extensive feature engineering or selection, we employ a deep learning classifier as the study's baseline classifier. Deep learning is a branch of artificial intelligence and currently the most successful computational intelligence technique for diagnosing ES in health informatics. This paper deals with a real-life application of epilepsy classification using computational techniques namely, Target-vs-One and Target-vs-All using deep learning approach. It is investigated that the baseline classifier on Target-vs-One strategy achieved the highest f1-score and accuracy about 0.9815 and 0.9818, respectively, as compared to the performance of baseline classifier on Target-vs-All strategy (e.g., achieved 0.94 of f1-score and 0.98 of accuracy).

Keywords: Deep learning · One-vs-One · One-vs-All · Epilepsy · Classification

1 Introduction

The neurological disorder epilepsy affects the brain and nerve system [1]. It is a well-known neurological condition that affects around 1 percent of the general population. Only in the United Kingdom, around 1 in 100, more than half a million people suffer

Á. Rocha et al. (Eds.): WorldCIST 2024, LNNS 986, pp. 85–94, 2024.
https://doi.org/10.1007/978-3-031-60218-4_9

from epilepsy [2]. According to a World Health Organization (WHO) estimate, ES ailments impact roughly 50 million people worldwide [3], or 5 to 10 people per 1000 people [4]. ES is a neurological disorder that needs the prediction of potentially life-threatening occurrences [5]. There are numerous potential causes of epilepsy, which are sometimes difficult to detect. In the field of epilepsy, seizure is known as an epileptic seizure, and the source is the brain. During an epileptic seizure, the regular functioning of the brain is temporarily disrupted, resulting in a breakdown of the signaling system between the brain and the rest of the body. These seizures can increase the likelihood that epilepsy patients would sustain common childhood traumas such as fractures, falls, burns, and submersion injuries [6]. It is considered that those with ES have a higher risk of accidental damage than those without ES. The increased risk may be caused directly by a seizure or indirectly by other comorbid disorders that enhance the likelihood of damage. Elaine C. Wirrell [6] summarized various studies [7–9] about various cases of ES into three important categories. First, seizures can cause falls that occur suddenly and without warning. The patient is unable to use their protective reflexes to arrest their fall, and as a result, they may get a head, orthopedic, or soft tissue damage. They may also fall onto a hot surface or into water, sustaining burns or submersion injuries. Second, absence or partial complex seizures result in loss of consciousness, preventing the patient from recognizing and responding to potentially hazardous situations. Even in the absence of clear clinical seizure activity, it has been demonstrated that paroxysmal EEG discharge affects alertness and mental speed [7]. Antiepileptic drugs may impair cognition, although this effect is probably modest in the majority of cases. While it is believed that newer drugs have less cognitive side effects, this may not be the case for all treatments [8]. Third, ES is known to be associated with a number of comorbid diseases that may play a role in raising injury risk. Attention deficit disorder is significantly more prevalent in children with epilepsy (37% vs. 5% in healthy controls) and has been linked to an increased risk of accidental harm [9]. These injuries occur due to the fact that a seizure can occur at any moment and in any location without notice, and the victim would continue their activity while unconscious. If a system can accurately forecast the pre-seizure phase (the time when the brain is transitioning towards generating a seizure), it could create an early warning signal so that the patient can take precautions.

Existing methods for ES detection that can be employed to extract characteristics from EEG signals include [10]: (i) useful for analyzing stationary signals, and (ii) capable of studying continuous EEG signal segments. However, due to the non-stationary nature of EEG signals, typical frequency analysis methods may not be able to capture all of the intricacies of brain signals. Traditional ML techniques can be distinguished from modern classification algorithms, also known as computationally intelligent algorithms or ML algorithms.

From the preceding discussion, it can be deduced that the advantage of using computational intelligence over traditional ML classification approaches is that the classification model does not need to know a great deal about the input data at the outset; rather, the input data's properties are learned by the computational intelligent learning technique. Once the classification system has been created and taught, it would not be necessary to modify the system in order to replace a learning method with another one. In light of this benefit, classifiers like deep learning have been widely employed in the development of ES detection and classification models. Therefore, we have selected deep learning as baseline classifier for ES classification in the real-world clinical epilepsy patients' data.

The rest of the paper is organized as follows. Section 2 explains the brief literature review and discusses the process of ES detection methods. It then explains computational intelligence techniques for combining the data for classification system such as one-vs-one and one-vs-all. Section 3 the proposed empirical framework and presents classification results, followed by results and discussion in Sect. 4. The paper concludes in Sect. 5.

2 Epilepsy Seizure and Computational Techniques

This section provides literature review of epilepsy seizures, ML techniques, and the deep learning strategy applied as the baseline classifier for ES.

2.1 Epilepsy Seizure (ES)

ES is a chronic, non-communicable brain ailment that can emerge for no apparent reason due to a genetic flaw. It affects people of all ages. The unpredictability of ES poses a serious threat to human life [11]. ES is a neurological disorder that needs the prediction of potentially life-threatening occurrences [5]. One of the basic epilepsy diagnostic tools is the electroencephalogram (EEG). Recognizing ES activity is frequently achieved by a human expert and is dependent on finding specific patterns in the multichannel EEG. However, this is a difficult and time-consuming procedure [12]. There are various time-consuming classical approaches for EEG signal analysis to detect ES exist [4]. The root cause of this condition is currently unknown. Despite this, many people can be medically treated for seizures if they are detected early.

2.2 Computational Techniques to Classify the ES

Typically, classification systems based on Bayesian statistical theory have been widely and successfully utilized. This method provides a means of representing sensory evidence, features retrieved from raw data, and prior knowledge about the task at hand gleaned from domain expertise. With identical prior probabilities for all classes and thus ignoring the bother of gaining domain knowledge, the analysis becomes quite simple [10]. Even though it is a powerful and straightforward rule to comprehend and implement, deriving posterior probabilities from data is a difficult operation, and data distribution may not be uniform.

Using long-term CHB-MIT EEG recordings, Dalton et al. [13] evaluated EEG signal temporal domain features as mean, standard deviation, zero-crossing rate, entropy, and root means square (RMS) to detect seizures. Their findings showed that the RMS is the best indicator of seizure. Average seizure categorization sensitivity and specificity were 91% and 84%.

According to Hussain L [14], SVMs are the most popular and effective machine learning classifier for seizures. For accurate epileptic seizure diagnosis, time-frequency analysis needs more investigation. It is vital to highlight that all existing machine learning algorithms are field-specific feature extraction methods. The unpredictability of seizures between patients is a problem for domain-based approaches. EEG data is non-stationary, and seizure statistics vary per patient. For the same patient, their vulnerability to artefacts fluctuates over time, which may reduce seizure detection systems' efficacy. Ayoubian et al. [15] used relative energy, number of peaks, and wavelet entropy to enhance seizure detection accuracy by 72% and false detection rate by 0.7 per hour. EEG features like sample entropy were used to identify epileptic episodes. After feeding the features to the extreme learning machine (ELM), classification accuracy was 95.67 percent, sensitivity was 97.26 percent, and specificity was 98.77 percent. Furthermore, Lee H. et al. [3] explored the use of MRI in identifying epileptogenic lesions through the application of machine learning techniques. It emphasized the use of automated algorithms for diagnosing and predicting the outcome of a medical condition.

Deep learning (DL) is a branch of ML that uses artificial neural networks (ANN) and simulation learning to create a model that equals the level of the human brain in addressing complicated issues in the real world [16, 17]. Each DL model must achieve two basic DL assignments tightly: extracting semantic information from input and providing meaningful output. Deep learning attempts to reach the level of these neural circuits by mimicking the human brain with ANN algorithms, which is at the heart of DL [18]. It has an input layer, an output layer, and one or more hidden layers. Each hidden layer comprises a group of weighted nodes or neurons connected to each other [17, 19]. On the other hands, Shoeibi A. et al. [1] investigated the utilization of DL in the detection of epileptic seizures utilizing neuroimaging. The text explores computer-aided diagnosis systems that are based on DL, as well as the datasets and models used in these systems. Further, they have encompassed rehabilitation techniques and evaluate research endeavors, tackling obstacles and suggesting future pathways for enhanced diagnosis and anticipation of seizures.

The unpredictability of ES poses a substantial risk to human life when a target variable with five seizure labels, namely pre-seizure, seizure, and seizure-free, is identified using real clinical data. We adopt a deep learning classifier as the study's baseline classifier in order to reliably classify seizure activity (e.g., the target label) without additional feature engineering or selection. Deep learning is the most successful computer intelligence technique for identifying ES in health informatics. It is a subfield of artificial intelligence. This study discusses a real-world application of classifying epilepsy using computational methodologies, particularly Target-vs-One (TvO) and Target-vs-All (TvA) applying a deep learning methodology.

3   Propose Framework and Evaluation Setup

In this section, we describe the proposed classification and evaluation framework for handling the ES problem as multi-class problem as well as binary class problem with deep learning networks.

3.1 Subject Dataset and Preprocessing

The subject dataset is gathered from the publicly available source[1] of the Bonn University and explored in order to get a basic understanding of its features [20]. The data sampling frequency was 173.61 Hz. However, the spectral bandwidth of the acquisition equipment, which ranges from 0.5 Hz to 85 Hz, is reflected in the time series.

During this phase, we have processed the raw dataset which contains five different TXT files. Table 1 provides the details of these files.

Table 1. Detail of the raw dataset's files

File Name			File Size	Samples
SET A	Z.zip with	Z000.txt - Z100.txt	564 kB	4096
SET B	O.zip with	O000.txt - O100.txt	611 kB	4096
SET C	N.zip with	N000.txt - N100.txt	560 kB	4096
SET D	F.zip with	F000.txt - F100.txt	569 kB	4096
SET E	S.zip with	S000.txt - S100.txt	747 kB	4096

Each sample has 4096 data points, which are divided into 23 data chunks of 178, and 1-s data points each. The 23 data chunks are then rearranged at random. Finally, 11,500 time-series EEG signal data samples for 500 patients are obtained. The last column "class label" shows 1, 2, 3, 4, 5 for the following five health conditions: 1) Epileptic Seizure Activity, 2) EEG recorded at tumor site, 3) EEG recorded in healthy brain area, 4) EEG recorded with closed eyes, and 5) EEG recorded with open eyes, respectively. Furthermore, the raw data is transformed into the CSV file format for the proposed study.

3.2 Model Evaluation Setup

Then, we provide a theoretical examination of the benefits of multi-class classification, a technique that allows us to classify test data into several class labels available in trained data as a model prediction. There are primarily two categories of multi-class categorization strategies such as (i) One-vs-one (OvO) technique divides a multi-class classification into a binary classification problem for each class., and (ii) One-vs-All (OvA) decomposes a multiclass classification into a binary classification issue for each pair of classes. However, we have focused on the target class label which is very important to accurately classify e.g., Epileptic Seizure Activity or 1 in the whole dataset. Therefore, we have constituted a multi-class classification strategy namely target-vs-one (TvO) and target-vs-all (TvA) instead of considering all the class labels and dealing equally with them using traditional classification strategies of OvA and OvO. The training dataset is used to train the baseline classifier (i.e., deep learning) with the following configuration as given in Table 2.

---

[1] URL: https://www.ukbonn.de/epileptologie/arbeitsgruppen/ag-lehnertz-neurophysik/downloads/ Last Access on 17 Oct, 2023 19:45.

Table 2. Configuration of baseline classifier

Parameters	Binary classification values	Multi-classification values
Epochs	50	50
Batch size	16	16
Activation function in inner layers	Relu	Relu
Activation function in output layers	Sigmoid	Softmax
Optimizer	Adam	Adam
Learning rate	0.001	0.001
Loss function	Binary cross entropy	Sparse categorical cross entropy

After the learning process, the classifier is evaluated using the testing set to evaluate its classification accuracy for unseen observations. The percentage of classification accuracy for both training and testing sets are calculated separately based on correctly classified observations within each individual TvO and TvA based generated datasets.

In order to evaluate the models during the experiments, we have used hold-out cross validation methods and calculated the performance using the following scientific evaluation measures. Table 3 describes the performance evaluation measures.

Table 3. Performance evaluation measures

Evaluation measures	Formula	Equation #
Recall	$TPR = TP/(TP + FN)$	(1)
Precision	$PPV = TP/(TP + FP)$	(2)
Accuracy	$ACC = (TP + TN)/(P + N)$	(3)
F1-Score	$F1 = 2TP/(2TP + FP + FN)$	(4)

3.3 TvO Strategy for ES Using Deep Learning

This is a heuristic strategy for multi-class classification that splits the dataset into one set for target class versus every other class. The target dataset has five different classes: '1', '2,' '3', '4' and '5'. This could be divided into five binary classification datasets, namely Target-vs-One as reflected in Table 4.

Table 4. Target versus every other one (TvO)

Subsets	Target class label	vs Every other one class
ES Subset-1	1	2
ES Subset-2	1	3
ES Subset-3	1	4
ES Subset-4	1	5

3.4 TvA Strategy for ES Using Deep Learning

This is a heuristic strategy for multi-class classification that splits the dataset into one dataset for target class versus all other classes. The subject dataset has a total of five different classes: '1,' '2,', '3', '4' and '5'. So, the first-class label '1' is the target label and grouped all other labels' data e.g., {2, 3, 4, 5}. This could be divided into one binary classification data set, namely Target-vs-One. The following Table 5 gives a summary of TvA for ES dataset classification.

Table 5. ES Target versus Other All (TvA)

Subsets	Target class label	vs Every other one class
ES Subset-5	1	{2,3,4,5}

4 Results and Discussion

In this section, the baseline classifier's performance evaluated between TvA and TvO based classification strategies. The model's performance of all the strategies is evaluated. Firstly, the experiments were performed using the original dataset without TvA and TvO data classification strategy. Table 6 shows the performance of the original dataset.

Table 6. Results of original dataset using deep learning

Evaluation measures	Results
Precision	0.77
Recall	0.77
Accuracy	0.76
F1-Score	0.76

Then, inconsequential experiment, the TvA strategy was adopted and evaluated using the classifier. The obtained performance of this experiment is presented in Table 7.

Table 7. Results of TvA using deep learning

Evaluation measures	Results
Precision	0.92
Recall	0.97
Accuracy	0.98
F1-Score	0.94

Finally, the TvO strategy tested using the baseline classifier and presented classification performance target label "T" in the following Table 8. The classification accuracy (%) of testing datasets (e.g., Tv2, Tv3, Tv5, and Tv5) is shown in Table 8 where "Average" represents the average performance for all four datasets distribution.

Table 8. Results of TvO using deep learning

Evaluation measures	Results				
	Tv2	Tv3	Tv4	Tv5	Average
Precision	0.9700	0.9888	0.9895	0.9895	0.9844
Recall	0.9763	0.9865	0.9772	0.9772	0.9793
Accuracy	0.9728	0.9880	0.9826	0.9826	0.9815
F1-Score	0.9731	0.9877	0.9833	0.9833	0.9818

The performance of the baseline classifier in terms of state-of-the-art evaluation metrics (e.g., accuracy and f1-score 0.76 or 76%) is quite poor when compared to the baseline classifier's performance when evaluating TvA or TvO data classification (see Table 6). Although TvA improves the performance of the baseline classifier, as demonstrated in Table 7, the performance of the baseline classifier is still superior. In contrast, the f1-score of the baseline classifier utilizing TvO is 0.98 as compared to f1-score 0.94 of baseline classifier on TvA. Therefore, the target-versus-one method with deep learning outperforms both the TvA strategy with and without data classification.

5   Conclusion

This study proposes and evaluates various multi-class epilepsy seizure classification strategies, including OvA, OvO, TvA, and TvO with supervised learning algorithms (e.g., deep learning). Using TvO, computationally sophisticated techniques for deep learning have proven useful to TvA at recognizing and classifying complex patterns in the input data (EEG signals). Since TvA's deep learning classifier performance was compared to that of TvO's deep learning classifier performance. In terms of accuracy, precision, recall, and f1-score, TvO with deep learning demonstrated the best recognition performance (e.g., achieved 0.9815 of accuracy, and 0.9818 of f1-score) as compared to the performance of baseline classifier on TvA (e.g., achieved 0.98 of accuracy and 0.94 of f1-score). As a result, it is investigated that the proposed data preprocessing strategies is suitable alternative approach for classification of the epilepsy seizure. Future research will study further categorization approaches based on sophisticated computationally intelligent techniques such as continuous learning, deep learning, and adaptive learning architecture in an effort to develop a method that is best in both noisy and noise-free situations. Furthermore, we will extend the spectral range to ensure comprehensive coverage of relevant frequencies and to assess the potential impact on the temporal coherence of EEG signals.

Acknowledgements. This work was supported by the FCT – Fundação para a Ciência e a Tecnologia, I.P. [Project UIDB/05105/2020].

References

1. Shoeibi, A., Moridian, P., Khodatars, M., et al.: An overview of deep learning techniques for epileptic seizures detection and prediction based on neuroimaging modalities: methods, challenges, and future works. Comput. Biol. Med. **149**, 106053 (2022). https://doi.org/10.1016/j.compbiomed.2022.106053
2. Qazi, K.I., Lam, H.K., Xiao, B., et al.: Classification of epilepsy using computational intelligence techniques. CAAI Trans Intell Technol **1**, 137–149 (2016). https://doi.org/10.1016/j.trit.2016.08.001
3. Lee, H.M., Gill, R.S., Bernasconi, N., Bernasconi, A.: Machine Learning in Neuroimaging of Epilepsy, pp 879–898 (2023)
4. Zubair, M., Belykh, M.V., Naik, M.U.K., et al.: Detection of epileptic seizures from EEG signals by combining dimensionality reduction algorithms with machine learning models. IEEE Sens. J. **21**, 16861–16869 (2021). https://doi.org/10.1109/JSEN.2021.3077578
5. Shiragapur, B., Dhope (Shendkar), T.S., Simunic, D., et al.: Predicting epilepsy seizures using machine learning and IoT. In: Smart Innovation of Web of Things. CRC Press, pp 63–82 (2020)
6. Wirrell, E.C.: Epilepsy-related Injuries. Epilepsia **47**, 79–86 (2006). https://doi.org/10.1111/j.1528-1167.2006.00666.x
7. Aldenkamp, A., Arends, J.: The relative influence of epileptic EEG discharges, short non-convulsive seizures, and type of epilepsy on cognitive function. Epilepsia **45**, 54–63 (2004). https://doi.org/10.1111/j.0013-9580.2004.33403.x
8. Elterman, R.D., Glauser, T.A., Wyllie, E., et al.: A double-blind, randomized trial of topiramate as adjunctive therapy for partial-onset seizures in children. Neurology **52**, 1338 (1999). https://doi.org/10.1212/WNL.52.7.1338
9. Dunn, D.W., Austin, J.K., Harezlak, J.: ADHD and epilepsy in childhood. Dev. Med. Child Neurol. **45**, 50–54 (2007). https://doi.org/10.1111/j.1469-8749.2003.tb00859.x
10. Jain, A.K., Duin, P.W., Mao, J.: Statistical pattern recognition: a review. IEEE Trans. Pattern Anal. Mach. Intell. **22**, 4–37 (2000). https://doi.org/10.1109/34.824819
11. Nanthini, K., Tamilarasi, A., Pyingkodi, M., et al.: Epileptic seizure detection and prediction using deep learning technique. In: 2022 International Conference on Computer Communication and Informatics (ICCCI). IEEE, pp 1–7 (2022)

12. Gramacki, A., Gramacki, J.: A deep learning framework for epileptic seizure detection based on neonatal EEG signals. Sci. Rep. **12**, 13010 (2022). https://doi.org/10.1038/s41598-022-15830-2

13. Dalton, A., Patel, S., Chowdhury, A.R., et al.: Development of a body sensor network to detect motor patterns of epileptic seizures. IEEE Trans. Biomed. Eng. **59**, 3204–3211 (2012). https://doi.org/10.1109/TBME.2012.2204990

14. Hussain, L.: Detecting epileptic seizure with different feature extracting strategies using robust machine learning classification techniques by applying advance parameter optimization approach. Cogn. Neurodyn. **12**, 271–294 (2018). https://doi.org/10.1007/s11571-018-9477-1

15. Ayoubian, L., Lacoma, H., Gotman, J.: Automatic seizure detection in SEEG using high frequency activities in wavelet domain. Med. Eng. Phys. **35**, 319–328 (2013). https://doi.org/10.1016/j.medengphy.2012.05.005

16. Aggarwal, K., Mijwil, M.M., et al.: Has the future started? The current growth of artificial intelligence, machine learning, and deep learning. Iraqi J. Comput. Sci. Math. **3**(1), 115–123 (2022). https://doi.org/10.52866/ijcsm.2022.01.01.013

17. Richards, B.A., Lillicrap, T.P., Beaudoin, P., et al.: A deep learning framework for neuroscience. Nat. Neurosci. **22**, 1761–1770 (2019). https://doi.org/10.1038/s41593-019-0520-2

18. Parisi, G.I., Kemker, R., Part, J.L., et al.: Continual lifelong learning with neural networks: a review. Neural Netw. **113**, 54–71 (2019). https://doi.org/10.1016/j.neunet.2019.01.012

19. Rammo, F.M., Al-Hamdani, M.N.: Detecting the speaker language using CNN deep learning algorithm. Iraqi J. Comput. Sci. Math. **3**(1), 43–52 (2022). https://doi.org/10.52866/ijcsm.2022.01.01.005

20. Andrzejak, R.G., Lehnertz, K., Mormann, F., et al.: Indications of nonlinear deterministic and finite-dimensional structures in time series of brain electrical activity: dependence on recording region and brain state. Phys. Rev. E **64**, 061907 (2001). https://doi.org/10.1103/PhysRevE.64.061907

Health Informatics

OralDentalSoft: Open-Source Web Application for Dental Office Management

Ricardo Burbano[iD], Eduardo Estévez[iD], Lucrecia Llerena[(⊠)][iD], and Nancy Rodríguez[iD]

Faculty of Engineering Sciences, State Technical University of Quevedo, Quevedo, Ecuador
{dburbano,ericksson.estevez2016,lllerena,nrodriguez}@uteq.edu.ec

Abstract. In Ecuador, numerous dental care facilities can be found, many of which are public. However, the organization of these centers has limitations. Recognizing the potential benefits that a specialized web application could offer to dental offices; we propose the development of such an application. We employ a case study approach as a research methodology to explore the development of an open-source web application in the field of dental health. The primary objective of this study is the creation of an open-source web application called "OralDentalSoft," designed to enhance the management of dental offices. We utilize the OSCRUM methodology for the development of this application to ensure efficiency and security. "OralDentalSoft" aims to improve the interaction between dental offices and their patients, providing a more efficient service. For this research, we conducted a systematic literature search. We highlight that, as of the date of this study, we have not found reports of open-source web applications specifically designed for dental offices. "OralDentalSoft" offers efficient record-keeping and control of information and activities related to dental health. User evaluation results indicate substantial improvements in information management and recording, optimization of the time spent on scheduling medical appointments easily and quickly, and more effective patient service. In conclusion, the web application developed using the methodology employed emerges as a beneficial tool for dental offices, enhancing the efficiency and effectiveness of their processes.

Keywords: Web Application · Dentist · Dental Health · Oral Health · OSCRUM

1 Introduction

Currently, web applications have deeply entrenched themselves in our daily lives, becoming ubiquitous tools that foster global interconnectivity [1]. One of their main advantages lies in accessibility, as they do not require a specific computer and allow any user with an internet-enabled device to use them [2]. In line with this technological evolution, the Ecuadorian government has embraced open-source software as an integral part of its technology policy. The adoption of open-source code not only promotes digital inclusion but also optimizes government spending [3]. The benefits of open-source software are diverse, including technological autonomy, standardization, security, vendor independence, democratization of information, cost-effectiveness, and health improvements [3].

© The Author(s), under exclusive license to Springer Nature Switzerland AG 2024
Á. Rocha et al. (Eds.): WorldCIST 2024, LNNS 986, pp. 97–106, 2024.
https://doi.org/10.1007/978-3-031-60218-4_10

Despite this widespread technological progress, in the field of healthcare, specifically in dentistry, manual patient and appointment records on paper persist, leading to administrative problems and affecting the efficiency of patient care, who lack access to a virtual record of their appointments [2]. In this context, the main objective of our research is to develop a web application that automates control in dental centers, facilitating more agile and efficient interaction between dentists and patients. The proposed open-source web application will expedite information automation, enabling efficient management of appointments and patients. This advancement aims to enhance communication between dentists and patients, increasing productivity in dental offices. The application will allow dentists to register patients, schedule appointments, and provide oral health advice, while patients will have access to a virtual dental record where they can keep track of their appointments, learn about updated dental service prices, and schedule their next oral cleaning appointment.

To address this research, a case study approach was applied, exploring the development of an open-source web application in the field of dental health. Additionally, a literature review was conducted through a Systematic Mapping Study (SMS) to identify articles related to the development of open-source projects in dentistry. This bibliographic analysis helped define the fundamental concepts necessary to create a high-quality web application accessible to any dental practice. The adopted development methodology, OSCRUM, an efficient variant of the agile SCRUM methodology, has proven valuable for gathering and understanding project requirements through business process modeling techniques [4]. This methodology promotes constant communication with the client, ensuring that the final product meets their expectations.

The open-source web application features an innovative approach that enables patients to directly interact with their dental history and appointment management. Additionally, it offers up-to-date information on dental services and pricing. The platform provides users with personalized oral care tips. This represents a significant shift in the dental care paradigm, granting patients greater control over their oral health and promoting active prevention.

The added value of our research lies in the freeness of the developed web application and its open-source nature, which implies significant savings for dental centers. Moreover, being open-source software, it can be tailored to the specific needs of each center, fostering participation from the open-source community for continuous improvements and updates. This innovative approach promotes collaboration and ensures a constant and beneficial evolution for the dental community. The document is organized as follows: Sect. 2 describes related work. Section 3 describes the OSCRUM methodology for the development of the OSS web application. Section 4 describes the case study. Section 5 outlines the case study results. Section 6 discusses the results, and finally, Sect. 7 presents the conclusions and future work.

2   Related Work

After conducting a literature search in the Scopus and Springer-Link databases, several papers reporting the use of open-source web applications in the healthcare field were found [5–16].

In several of the related articles, different aspects of open-source platforms and digital health tools are explored. For example, the studies by Huang [5], Jäckle [6], and Wadali [7] focus on the development and evaluation of open-source platforms for data collection in the healthcare domain. Huang concentrates on creating a versatile platform, while Jäckle focuses on a platform-specific to virus propagation. On the other hand, Wadali evaluates open-source DICOM viewers for India's National Telemedicine Service. These studies highlight the importance of open-source tools in healthcare data collection, although Jäckle stands out for using an agile development methodology in contrast to the lack of methodology specification in Huang and Wadali's studies.

Meanwhile, the studies by Stefan Konigorski [8] and Kotoulas [9] center on the realm of clinical trials. Konigorski presents an open-source platform for N-of-1 trials, used to assess the effectiveness of specific treatments in individual patients. In contrast, Kotoulas focuses on developing a biomedical registry framework to improve the efficiency of clinical trials. Despite the relevance of these approaches, none of the studies specify the development methodology used in creating these platforms.

The papers by Zucker [10] and Samal [11] address the topic of open-source tools for data analysis and visualization in healthcare. Zucker introduces a web application for routine data collection and management, while Samal focuses on an automated clinical decision support tool for chronic kidney disease. However, neither of the studies provides detailed information about the development methodology of the tools. In the realm of genomic and clinical data management, the studies by Fasemore [12] and Glicksberg [13] are relevant. Fasemore presents CoxBase, a platform for epidemiological surveillance and genomic sequence analysis, while Glicksberg uses blockchain technology to share genomic and clinical data of cancer patients. In both cases, the development methodology of the tools is not specified.

The study by Seid [14] focuses on developing an agent-based model for collaborative learning in healthcare, where patients, healthcare providers, and researchers collaborate to enhance healthcare. Despite its relevance, the author does not mention the development methodology used. Additionally, the article by Dong [15] introduces a tool for normalizing COVID-19 test names to LOINC codes. Unlike other studies, Dong specifies the use of agile development methodology in creating this tool. Finally, the study by Tom-Aba [16] evaluates the maturity of SORMAS, a digital health platform, and mentions the use of agile development methodology in conducting this evaluation. In conclusion, none of the works identified during the literature review employ a specific development methodology for creating an open-source web application in the dental health domain. Therefore, it has been established that this research study will serve as a foundation for future investigations in this field.

3 Description of the Proposed Methodology for Open-Source Development: OSCRUM

In this section, the OSCRUM methodology for open-source software development is presented. OSCRUM is a variant of SCRUM that includes a way to gather requirements quickly and simply. In the case of "OralDentalSoft," OSCRUM has been used to ensure agile and effective development of the web application. This methodology is based on three main pillars: transparency, testing, and adaptation. It also follows five values that typically guide the development process: personal commitment, open-mindedness, high competency level, goal-oriented focus, and collaboration. The development of the open-source web application OralDentalSoft considers the framework of the OSCRUM methodology through 11 activities outlined according to Rahman et al. [17] (See Table 1). For the construction of the web application, several open-source tools have been used, such as Angular, Firebase, Bootstrap, TypeScript and Visual Studio Code.

Table 1. OSCRUM Activities for Web System Development [17]

No	CRITERION	DEFINITION
1	Discovery of the Problem and Volunteer Search	Information is gathered through web artifacts, and brainstorming sessions with a small group of individuals
2	Communication	A meeting is held to list the features that the software will have
3	Initial Launch Planning Meeting	Meetings are scheduled to work on the initial functionalities and the corresponding Sprint tasks
4	Product Launch Plan and Status	Once the Sprints are completed, the characteristics are listed with their respective status
5	Feature Update	If any new feature is approved after listing the Sprint, will be updated
6	Test the source code	The users and the general community can test the source code within the GitHub repository
7	Report errors	The users can report any issues or problems they encounter
8	Contributions from external members	Acceptance of individuals outside the project so that they can collaborate on new features for the system
9	Repair	The problems or errors previously encountered will be corrected
10	Approval	After resolving the above, a new evaluation is carried out to determine if the system has any errors; if it does not, approval is granted
11	Activity iteration	The process will be repeated as many times as necessary to meet the product quality standards

4 Case Study

This section presents the development of the web system 'OralDentalSoft' using the OSCRUM methodology. Below, we detail the activities carried out with the OSCRUM methodology for the development of the open-source project OralDentalSoft.

Problem Discovery and Volunteer Recruitment. The OSCRUM methodology allowed researchers to establish a starting point for the development of the open-source system. The first step was to identify the problem and recruit volunteers. For this purpose, a web artifact (blog) was developed where the main idea of the software project was shared. This blog can be accessed through the following link: https://construcc ionsoftwaresalud.blogspot.com/. The objective of this blog was to gather information from the open-source community for their participation in tests, such as their names and email addresses. Additionally, it was used as a communication channel between the development team and the community that will use this system.

Communication. The requirements for the implementation of OralDentalSoft, both functional and non-functional, are detailed in this section. These requirements were obtained through various techniques such as brainstorming and web tool analysis (competitor analysis). To generate brainstorming, a virtual meeting was held on Google Meet, where all participants contributed their ideas and suggestions. Additionally, a blog was used to collect comments and suggestions from experts in the field, such as dentists, and a competitive analysis was conducted. The result of applying these techniques, including the blog, is presented in a document detailing the requirements and is available at the following link: https://drive.google.com/file/d/1FmRr5UtnpGjequMn2gm1TXYiFjaO1 agF/view.

Initial Release Planning Meeting. After the initial communication meeting, a second meeting was held to plan the initial release of OralDentalSoft. During this meeting, all the functional and non-functional requirements necessary for the successful implementation of the project were identified and defined. Effective information-gathering techniques such as brainstorming and competitive analysis were used for this task. Google Docs was the technology tool used for sprint planning, allowing real-time collaboration and access to documents from anywhere. Thanks to the use of these techniques and tools, a clear definition of the project requirements was achieved. For information on the technical requirements of OralDentalSoft, you can refer to the initial requirements report at the following link: https://drive.google.com/file/d/12wH64Dmcpaqyx4mFDqMS5mUlZY BgsjnW/view. Additionally, to learn how brainstorming and competitive analysis were applied in the requirements definition process, you can access the following page: https://drive.google.com/file/d/1tt3iTOIR5ytl7mKUQDSMEH5SL1KHLu6m/view.

Release Plan and Product Status. Once the sprint backlog of features was completed, a release plan was developed, including a list of activities and their current status. Users can test the system in production through the link available on our blog. The OralDentalSoft release plan details all the functionalities and features of the system, along with their current development status. Progress information on backlog features is available for reference on the blog(https://construccionsoftwaresalud.blogspot.com/).

Feature Updates. In this activity, a review of the features provided by the community within the blog was conducted. This information was accepted by the main maintainer, and once the features were accepted, work on updating the backlog began. It's worth noting that pilot tests were conducted with friends and/or acquaintances of the researchers (seventh-semester students at the Technical State University of Quevedo). These users reviewed the blog and provided valuable feedback on the prototypes and project features.

Testing the Source Code. This activity involved running tests on the source code to identify and correct errors, bugs, and compatibility issues. To carry out this activity, the project code was uploaded to a GitHub repository, allowing each collaborator or user to test the functionalities, and the maintainer could review the code to be uploaded to the repository. The project code can be found at the following GitHub repository link: https://github.com/eduardo96452/Proyecto_OralDental. To recruit external users to participate in these tests, invitations were sent to friends and colleagues of the researchers to access our blog. Only interested external users provided comments and left their email addresses for contact. Once the user's email was known, resources were sent to proceed with testing the source code and error reporting. The results of these tests were documented and used to improve the system. Additionally, a service from the Firebase platform named Hosting was used. You can find the OralDentalSoft application at the following link: https://oraldentalsoft.web.app/.

Reporting Errors. In this activity, a template was designed to list the errors reported by users in the web application. Users were invited to report any errors they encountered while using the system. The development team was responsible for reviewing and documenting the reported errors, creating a list available at the following link: https://drive.google.com/file/d/1n7iemHwNhGmPsER9gnpknuatNjX52jvU/view.

Contributions from External Members to the Project. In OralDentalSoft, the open-source community contributed their knowledge and skills to the development of the system through an open participation activity. To facilitate this contribution, an open communication channel, such as a blog, was created where external users shared their ideas, suggestions, and solutions. Necessary documentation was provided, and technical support was offered to external users to contribute to the project. All contributions received were reviewed and evaluated by the development team to ensure they met the required quality standards. This way, the system was continuously improved, providing a better user experience. Valuable suggestions for improving the system were obtained through this open collaboration. Some of these suggestions included removing buttons with undefined interfaces, button redirection, data input validation, adding confirmation feedback for the user, improving the login interface, and correcting links that directed to the application's homepage. These suggestions were implemented thanks to error reports received from users who participated in the tests, ultimately improving the quality of OralDentalSoft.

Error Fixing. This activity was carried out by the OralDentalSoft development team and was reported to the open-source community. Once the problem to be resolved was identified, it was assigned to a team member to address and verify its functionality before being implemented into the source code. This activity was crucial to ensure the quality and stability of the project and is a key element in the ongoing system repair and improvement process.

Approval. During this activity, the development team leader reviewed the work that was done and determined if it met the established standards and requirements before final approval and release to the public. Approval was critical to guarantee the quality of the OralDentalSoft open-source code project.

Iteration. Within the system development process, multiple iterations have been applied to implement all desired features in the OralDentalSoft web system.

5   Case Study Results

In this section, we present the results derived from the application of the OSCRUM methodology [17] in the development of the open-source web application OralDental-Soft, detailing its implementation and access to its source code. These results are based on the requirements previously established in the OralDentalSoft feature list, which includes crucial aspects such as dentist registration, patient management, appointment scheduling, dental history, and viewing upcoming appointments, among others. Through this approach, we have successfully created a web application designed for the effective management of a dental clinic, enhancing the quality and automation of the services provided by dentists. OralDentalSoft has been designed to cater to different types of users: (i) Administrator User, responsible for creating and updating dentist accounts, (ii) Dentist User, with access to functionalities including logging in, viewing the patient list, creating new patients, scheduling appointments, and reviewing dental histories, and (iii) Patient User, with the ability to log in, schedule appointments, and review their dental history.

A meeting was held to coordinate the initial launch of OralDentalSoft, during which priorities for each functionality specified in the software feature list were defined. During this meeting, the implementation of each feature was planned, and the necessary tasks for its development were outlined. Regarding feature updates, no new functionalities have been added to date, as no additional needs were identified in the user community. However, flexibility is maintained to incorporate new features if future demands arise and are approved by the project's primary responsible party after analysis to determine their viability and scalable integration with the system. The web application uses an HTTPS encryption security system and ensures proper authorization to restrict access to sensitive data. To achieve this, a system of roles and permissions is employed that limits the information available to each user. Subsequently, after the community evaluated OralDentalSoft deployed on the server, error reports were recorded using templates, without significant difficulties in its usage. The final phase of the OSCRUM methodology (iteration), is repeated as needed to add new features to the software project and maintain existing functionalities.

OralDentalSoft is an open-source web application developed to automate the management of dental clinics. The project's source code is hosted in a GitHub repository, allowing any interested user to access it and collaborate on its development and contribute improvements. Additionally, this link provides access to documentation, development details, and changes made to OralDentalSoft. To download the source code, the following link is used: https://github.com/eduardo96452/Proyecto_OralDental. Figure 1 presents the main interface of OralDentalSoft.

6 Discussion of Results

The OralDentalSoft web application was conceived to manage dental clinics and facilitate interaction between dentists and patients. In addition to its functionality, it stands out for its versatility, as it is accessible from various devices, providing a high degree of convenience. It is important to highlight the relevance of the OSCRUM methodology in the development of this software, as it played a fundamental role in the planning and creation of a high-quality system. Consequently, it has been demonstrated that the OSCRUM methodology is a highly effective tool for the development of open-source projects, providing a clear and organized structure that has led to satisfactory results.

During the course of this research, some notable limitations were identified. Primarily, there was a lack of detailed information on specific development methodologies for open-source projects. While numerous documents describing web applications related to health were found and significantly contributed to the research, they did not delve into the implementation of methodologies for open-source project development. Second, the sample used in the study was limited to patients from a single dental clinic in an urban area, which could limit the generalization of results to other geographical and socioeconomic populations. Patients with certain medical conditions were also excluded, and it was observed that patients in the control group did not receive specific instructions on improving their oral hygiene.

Concerning the case study, several limitations related to the validity and reliability of its contributions are acknowledged. Although no issues with construct validity and internal validity were detected, the main limitation of our study lies in the number of case studies conducted (only one OSS development). Therefore, it is suggested to carry out additional case studies to apply the OSCRUM methodology to other OSS projects and thereby enhance external validity. Furthermore, it is considered essential to expand the participation of OSS users in future case studies to ensure a more comprehensive and reliable representation of all types of OSS users. Therefore, exploring other alternatives, such as social media, to recruit users interested in participating in such research is recommended.

Fig. 1. Main Interface of the OralDentalSoft Web Application.

7   Conclusion and Future Work

The fundamental purpose of this study is the creation of OralDentalSoft, an open-source web application, with the primary goal of improving the efficiency and quality of health-care in the field of dentistry by enhancing the management of dental clinics. The activities carried out for its implementation have played a crucial role in the research and development of the web application, allowing for the identification and adoption of a methodology specially tailored to the needs of this project. The result of this effort is software that meets the demands and expectations of the open-source community. The implementation of the OSCRUM methodology has been instrumental in developing software that satisfies user needs, is user-friendly, and is effective in managing and creating patient appointments and medical histories. The comprehensive review of the literature has been essential to utilize relevant information, including user contributions in repositories and web artifacts. From this review, two crucial aspects stand out: (i) the creation of web artifacts to keep the community informed and (ii) the use of repositories to host the source code, optimizing collaboration and transparency in the development of the application. Ultimately, it is expected that this OralDentalSoft web application will significantly contribute to the improvement of dental health for patients. With that purpose in mind, it is suggested that future research delve deeper into understanding its impact and effectiveness in various populations and clinical contexts. Additionally, the incorporation of new functionalities into the application, such as brushing reminders and the ability to send monitoring data to dentists or caregivers, is recommended to maximize its utility and benefit in the field of dental health.

References

1. Domínguez, J.D.T., Rodríguez, A.E.T., Danilo Valdés Ramírez: Tendencias en el desarrollo de aplicaciones web. In: XVI Convención Científica de ingeniería y Arquitectura (2012)
2. Mero Lino, E.A., Ramirez Cevallos, L.D.: Implementación de aplicación web para la gestión de proceso en el área de odontología del departamento de bienestar universitario de la Universidad Estatal del Sur de Manabí., (2022)
3. Moreno Garcés Lenín: Decreto 1017. 16 De Marzo. 1–28 (2020)
4. Rahman, S., et al.: OSCRUM : A Modified Scrum for Open Source Software Development, 1–7. https://doi.org/10.5013/IJSSST.a.19.03.20
5. Huang, H., et al.: A versatile and scalable platform that streamlines data collection for patient-centered studies: usability and feasibility study. JMIR Formative Res. 6, e38579 (2022). https://doi.org/10.2196/38579
6. Jäckle, S., Alpers, R., Kühne, L., Schumacher, J., Geisler, B., Westphal, M.: EsteR - a digital toolkit for COVID-19 decision support in local health authorities. Stud. Health Technol. Inf. 296, 17–24 (2022). https://doi.org/10.3233/SHTI220799
7. Wadali, J.S., Sood, S.P., Kaushish, R., Syed-Abdul, S., Khosla, P.K., Bhatia, M.: Evaluation of free, open-source, web-based DICOM viewers for the Indian national telemedicine service (eSanjeevani). J. Digit. Imaging 33, 1499–1513 (2020). https://doi.org/10.1007/s10278-020-00368-4
8. Konigorski, S., et al.: StudyU: a platform for designing and conducting innovative digital N-of-1 trials. J. Med. Internet Res. 24, e35884 (2022). https://doi.org/10.2196/35884

9. Kotoulas, A., Lambrou, G., Koutsouris, D.D.: Design and virtual implementation of a biomedical registry framework for the enhancement of clinical trials: colorectal cancer example. BMJ Health and Care Inform. **26**, 1–10 (2019). https://doi.org/10.1136/BMJHCI-2019-100008

10. Zucker, K., Wagstaff, M., Tomson, C., Beecham, R., Hall, G.: AuguR: A Scalable Open-Source Interactive Web Application for Routinely Collected Data. (2022). https://doi.org/10.3233/SHTI220177

11. Samal, L., D'Amore, J.D., Bates, D.W., Wright, A.: Implementation of a scalable, web-based, automated clinical decision support risk-prediction tool for chronic kidney disease using C-CDA and application programming interfaces. J. Am. Med. Inf. Assoc. JAMIA. **24**, 1111–1115 (2017). https://doi.org/10.1093/jamia/ocx065

12. Fasemore, A.M., et al.: CoxBase: an online platform for epidemiological surveillance, visualization, analysis, and typing of coxiella burnetii genomic sequences. mSystems **6**, e0040321 (2021). https://doi.org/10.1128/mSystems.00403-21

13. Glicksberg, B.S., et al.: Blockchain-authenticated sharing of genomic and clinical outcomes data of patients with cancer: a prospective cohort study. J. Med. Internet Res. **22**, e16810 (2020). https://doi.org/10.2196/16810

14. Seid, M., Bridgeland, D., Bridgeland, A., Hartley, D.M.: A collaborative learning health system agent-based model: computational and face validity. Learn. Health Syst. **5**, e10261 (2021). https://doi.org/10.1002/lrh2.10261

15. Dong, X., et al.: COVID-19 TestNorm: a tool to normalize COVID-19 testing names to LOINC codes. J. Am. Med. Inf. Assoc. JAMIA. **27**, 1437–1442 (2020). https://doi.org/10.1093/jamia/ocaa145

16. Tom-Aba, D., et al.: The surveillance outbreak response management and analysis system (SORMAS): Digital health global goods maturity assessment. JMIR Public Health Surveill. **6**, 1–9 (2020). https://doi.org/10.2196/15860

17. Rahman, S.S.M.M., et al.: OSCRUM: A Modified Scrum for Open Source Software Development. (1)AD. https://doi.org/10.5013/IJSSST.a.19.03.20

A Scoping Review of the Use of Blockchain and Machine Learning in Medical Imaging Applications

João Pavão[1] , Rute Bastardo[2] , and Nelson Pacheco Rocha[3]([⊠])

[1] Science and Technology School, INESC-TEC, University of Trás-Os-Montes and Alto Douro, Vila Real, Portugal
jpavao@utad.pt
[2] Science and Technology School, UNIDCOM, University of Trás-Os-Montes and Alto Douro, Vila Real, Portugal
[3] Department of Medical Sciences, IEETA, University of Aveiro, Aveiro, Portugal
npr@ua.pt

Abstract. This scoping review systematizes the current research related to the use of both blockchain and machine learning techniques in medical imaging applications. A systematic electronic search was performed, and twenty-five studies were included in the review. These studies aimed to use blockchain and machine learning techniques to provide (i) efficient security mechanisms to support the communication of medical imaging data, (ii) aggregation of distributed medical imaging data to train machine learning algorithms, and (iii) machine learning algorithms based on federated learning strategies. Among the ten machine learning techniques identified in the included studies, Convolutional Neural Network was the most representative (i.e., 44% of the studies). Moreover, Artificial Neural Network, Capsule Network, Deep Neural Network, Gated Recurrent Units, and Neural Network were machine learning techniques used by more than one study. Although the included studies developed algorithms with potential impact in clinical practice, it must be noted that they did not discuss the generalizability of their algorithms in real-world clinical conditions.

Keywords: Medical imaging · Blockchain · Machine learning · Scoping review

1 Introduction

Machine learning techniques are being extensively used in medical applications, namely, to support patients' diagnosis and prognosis, drug discovery and development or classification of clinical documentation using natural languages [1]. Especially in the field of medical imaging, in recent years, machine learning algorithms have been developed to improve the accuracy of diagnosis and prognosis [2], which is extensively documented by scientific literature (e.g., [3–5]).

Á. Rocha et al. (Eds.): WorldCIST 2024, LNNS 986, pp. 107–117, 2024.
https://doi.org/10.1007/978-3-031-60218-4_11

Moreover, machine learning techniques might be used together with blockchain (i.e., a cryptographically secure chaining of information blocks [6]) not only to protect the clinical data required for the training and execution of machine learning algorithms, but also to provide innovative ways to protect clinical data at transit or at storage [7], as being argued by different authors (e.g., [8–12]).

Therefore, it is important to analyse the research related to the use of both blockchain and machine learning techniques in medical imaging applications. In this respect, the present scoping review aimed to systematize state-of-the-art studies using blockchain and machine learning techniques to improve patients' diagnosis and prognosis supported on medical imaging data from different modalities, such as histopathology, ultrasounds, X-ray, computer tomography (CT), magnetic resonance imaging (MRI), positron emission tomography (PET), or single-photon emission computed tomography (SPECT). After this first section with the introduction, the rest of the paper is organized as follows: (i) the methods that were applied; (ii) the respective results; and (iii) a discussion and conclusion.

2 Methods

Considering the aforementioned objective, a review protocol was defined that is briefly described in this section.

The authors planned to search PubMed, Scopus, and Web of Science for references published before the end of September 2023. The expression defined for the electronic search was: ('medical image' OR 'medical imaging' OR 'radiology' OR 'computerized tomography' OR 'magnetic resonance') AND ('blockchain' OR 'block chain' OR 'distributed ledger'). This expression was complemented with search filters to exclude conference proceedings.

Table 1 presents the eligibility criteria to be used to achieve the list of the studies to include in the scoping review.

Based on demographic data, the authors planned a synthesis of the characteristics of the included studies, considering (i) the distribution of the studies by publication years, and (ii) the distribution of the studies by geographical areas, considering the affiliation of the first authors.

Moreover, it was considered a classification of the included studies according to their purposes and a characterization of their experimental setups (i.e., studies' specific objectives, medical imaging modalities, machine learning techniques being used, experimental datasets, and results).

Table 1. Eligibility criteria.

Inclusion	Exclusion
English references published in peer reviewed journals	Non-English references or references not published in peer reviewed journals
References reporting evidence of the use of blockchain and machine learning in medical imaging applications	References not related to the specific objective of this scoping review
References published before 30th September 2023	References published after 30th September 2023
References reporting on primary studies (i.e., original research)	References reporting on non-primary studies, including editorials, position papers, literature reviews or surveys
	References whose full texts were not available
	References reporting primary studies already reported by other references included in the revision. In this case, the reference reporting the most mature version of the study was selected

3 Results

3.1 Studies' Selection

Figure 1 presents the flowchart of the scoping review. The databases search was conducted in the first week of October 2023, and 479 references were retrieved: (i) 38 references from PubMed; (ii) 360 references from Scopus; and (iii) 181 references from Web of Science.

The initial step of the screening phase yielded 229 references by removing duplicates (n = 173), reviews (n = 35), books or book chapters (n = 36), and references without abstracts or authors (n = 6).

Based on titles and abstracts, 197 references were removed since they did not meet the eligibility criteria. Finally, the full texts of the remaining 32 references were screened and seven were excluded: (i) non-relevant use of blockchain, three references; (ii) retraction, two references; (iii) not written in English, one reference; and (iv) the experimental study was reported by another reference included in the review, one reference. Therefore, twenty-five studies were considered eligible and included in the review [13–37].

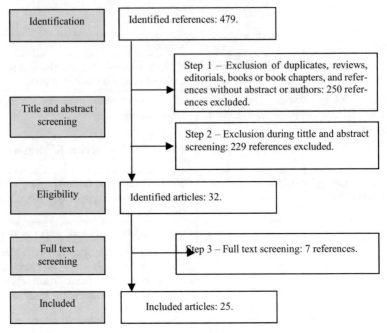

Fig. 1. Review flowchart.

3.2 Demographic Characteristics

The included studies were published between 2020 and 2023: (i) 2020, one study [13]; (ii) 2021, two studies [14, 15]; (iii) 2022, six studies [16–21]; and (iv) 2023, sixteen studies [22–37].

Concerning the geographical distribution, the most representative countries were China, six studies [14, 15, 19, 35–37], India, six studies [17, 21, 26, 29, 32, 33], and Saudi Arabia, six studies [16, 23–25, 28, 30], followed by Iran, two studies [18, 27]. Moreover, six countries contributed with one study each: Pakistan [20], South Korea [34], The Netherlands [13], United Arab Emirates [31], and United Kingdom [22].

3.3 Purposes of the Studies

Although the included studies developed and evaluated machine learning algorithms to surpass specific clinical problems, none of them presented a fully functional system to be used in clinical practice. Based on the analysis of the identified problems and the proposed solutions, the purposes of the included studies were divided into three classes (Table 2):

- Secure communication. The first subset of studies used data encryption, blockchain and machine learning techniques to design efficient security mechanisms to support the communication of medical imaging data.
- Data aggregation. Large high-quality datasets are required to train machine learning algorithms. The aggregation of clinical data from distributed points of care has the

potential to improve machine learning algorithms and, consequently, their diagnostic efficiency and accuracy, but decreases the security of the clinical data sources. Therefore, a subset of the studies was focused on the use of blockchain to secure the aggregation of medical imaging data stored in decentralized systems.

- Federated learning. Federated learning [38] allows data from locally trained machine learning models to be collected to create a more accurate global model. Therefore, federated learning does not require the sharing of patients' clinical data, but it still requires the sharing of other forms data such as parameters of locally trained machine learning models (e.g., model gradients or weights). In this respect, the purpose of the last subset of studies (i.e., federated learning) was to use blockchain to implement decentralized architectures allowing trusted multi-party machine learning algorithms for medical imaging applications without the need of the mediation of centralized trust servers [13].

Table 2. Purposes of the studies.

Purposes	Number of studies	Studies
Secure communication	2	[16, 35]
Data aggregation	13	[17, 18, 20–26, 28, 30, 32, 36]
Federated learning	10	[13–15, 19, 27, 29, 31, 33, 34, 37]

3.4 Experimental Characteristics of the Studies

Table 3 presents the characteristics of the two studies proposing blockchain and machine learning techniques to reinforce encryption and decryption to guarantee secure communications of medical imaging data using insecure channels (e.g., public communication channels).

Table 3. Secure communication.

Objective	Studies	Modality	Machine learning	Dataset	Results
Encryption and decryption	[16]	–	DNN	–	(1)
Encryption and decryption	[35]	MRI, PET, SPECT	CNN	Harvard Brain Atlas	(2)

(1) Without evaluation; (2) Comparison of the proposed algorithm with other algorithms described in scientific literature

The machine learning techniques that were applied were Deep Neural Network (DNN) [16] and Convolutional Neural Network (CNN) [35], but one of the studies [16] did not perform the evaluation of the proposed algorithm. The second study [35] evaluated the respective algorithm comparing its performance with the performance of other algorithms described in scientific literature. For that, the study [35] used the Harvard Brain Atlas dataset, which contains MRI, PET, and SPECT images.

Considering the thirteen studies classified as data aggregation (Table 4), blockchain was used to implement distributed mechanisms for securing medical images required by machine learning algorithms with the following objectives: (i) analysis of skin lesions images [23]; ii) brain tumor prediction [28]; (iii) breast cancer detection [24, 26]; (iv) classification of synovial sarcoma cancer [17]; (v) COVID-19 detection [18, 22]; (vi) images classification [30]; (vii) lung cancer prediction [21]; (viii) monkeypox detection [25]; (ix) osteosarcoma cancer detection [20]; (x) pneumonia image classification [36]; and (xi) skin lesion diagnosis [32].

Two studies [23, 32] did not refer to any medical imaging modality. In turn, two studies were focused on X-ray [22, 36], three studies on CT [18, 21, 30], four studies on histopathology [17, 20, 24, 26], one study on MRI [28] and one study on skin images [25].

This subset of studies applied the following machine learning techniques: (i) Artificial Neural Network (ANN) [23]; (ii) CNN [18, 20–22, 25, 28, 30]; (iii) Fully Convolutional Network (FCN) [32]; (iv) Gated Recurrent Units (GRU) [24, 26]; (v) Kernel Extreme Learning Machine (KELM) [32]; (vi) Neural Network (NN) [36]; and (vii) Support Vector Machine (SVM) [17];

Moreover, in terms of the results of the experimental setups, six studies [21, 22, 24–26, 32] performed the validation of the proposed algorithms, five studies [17, 18, 20, 24, 36] evaluated algorithms' performance, and three studies [23, 28, 30] compared the proposed algorithms with other algorithms described in scientific literature. Three studies [17, 21, 22] did not indicate the datasets used to obtain the experimental results, one study [32] used a benchmark skin dataset [32], one study [30] used several brain CT images comprising five familiar brain disorders [30], and one study [18] used a local dataset composed of clinical information of five Indian hospitals. The remainder seven studies used publicly available medical datasets: (i) COVID-19 posterior-anterior chest X-ray [36]; (ii) Hematoxylin and eosin-stained osteosarcoma histology images (H&E) [20]; (iii) Monkeypox-Dataset [25]; (iv) skin images from the International Skin Imaging Collaboration (ISIC) database [23]; (v) Wisconsin Diagnostic Breast Cancer (WDBC) [24, 26]; and (vi) Brain tumor classification MRI scan dataset available on Kaggle [28].

Table 5 presents the experimental characteristics of the eleven studies focused on federated learning approaches. Some proposed solutions (i.e., [19, 29, 31, 33, 34, 37]) are generic, while other studies implemented federated learning algorithms focused on specific objectives: (i) prediction of lung cancer survival [13]; (ii) COVID-19 detection [14]; and (iii) lung cancer detection [15, 27]. Moreover, in terms of medical imaging modalities, one study [31] did not refer to any medical imaging modality. In turn, most studies were focused on CT [13–15, 19, 27, 29]. The remainder studies were focused on CT and X-ray [34], and ultrasounds [37].

Table 4. Data aggregation.

Objective	Studies	Modality	Machine learning	Dataset	Results
Classification of synovial sarcoma cancer	[17]	Histopathology	SVM	–	(2)
COVID-19 detection	[18]	CT	CNN	Local	(2)
Osteosarcoma cancer detection	[20]	Histopathology	CNN	H&E	(2)
Lung cancer prediction	[21]	CT	CNN	–	(1)
COVID-19 detection	[22]	X-ray	CNN	–	(1)
Analysis of skin lesions images	[23]	–	ANN	ISIC	(3)
Breast cancer detection	[24]	Histopathology	GRU	WDBC	(1)
Monkeypox detection	[25]	Skin Images	CNN	Monkeypox	(1)
Breast cancer detection	[26]	Histopathology	GRU	WDBC	(1)
Brain tumor prediction	[28]	MRI	CNN	Kaggle	(3)
Images classification	[30]	CT	CNN	Images	(3)
Skin lesion diagnosis	[32]	–	FCN, KELM	Skin dataset	(1)
Pneumonia image classification	[36]	X-ray	NN	COVID-19	(2)

(1) Algorithm validation; (2) Algorithm performance; (3) Comparison of the proposed algorithm with other algorithms described in scientific literature

The subset of the included studies applied a set of machine learning techniques: (i) Capsule Network (CN) [14, 27]; (ii) CNN [13, 14, 34, 37]; (iii) DNN [15]; (iv) NN [29, 33]; and (v) Recurrent Convolutional Neural Network (RCNN) [15]. One study [19] did not refer to the machine learning technique being applied.

One study [33] included an experimental setup to compare proposed federated learning approach with centralized learning approaches. The experimental setups of the remainder studies [13–15, 19, 27, 31, 33, 34, 37] aimed to evaluate algorithms' performance, namely the performance of the applied machine learning techniques. For that, one study [13] used the NSCLC-Radiomics dataset, four studies [19, 27, 31, 34] used

Table 5. Federated learning.

Objective	Studies	Modality	Machine learning	Dataset	Results
Lung cancer prediction	[13]	CT	CNN	NSCLC	(1)
COVID-19 detection	[14]	CT	CN	Local	(1)
Lung cancer detection	[15]	CT	RCNN, DNN	Local	(1)
–	[19]	CT	–	Multiple	(1)
Lung cancer detection	[27]	CT	CN	Multiple	(1)
–	[29]	CT	NN	Local	(1)
–	[31]	–	ANN	Multiple	(1)
–	[33]	CT	NN	Local	(2)
–	[34]	CT, X-ray	CNN	Multiple	(1)
–	[37]	Ultrasounds	CNN	Local	(1)

(1) Algorithm performance; (2) Comparison with centralized learning

multiple datasets (i.e., Dataverse Harvard Repository and local datasets from six Chinese hospitals [19], Imaging Archive, Kaggle Data Science Bowl, LUN 16 and a local data set [27], Breast Cancer Wisconsin (Diagnostic) Dataset, Lung Cancer Prediction Dataset and Diabetes UCI Dataset [31], a COVID-19 X-ray and normal X-ray publicly available datasets [34]), and five studies [14, 15, 29, 33, 37] used local datasets (i.e., CT-scans collected from 89 patients of three different Chinese hospitals [14], 5842 CT cases [15], 34000 COVID-19 CT-scans from three different Indian hospitals [29], Chest CT-Scans [33], and a dataset of Beijing Children's Hospital [37]).

4 Discussion and Conclusion

This scoping review identified twenty-five studies using data encryption, blockchain and machine learning techniques to support medical imaging applications. The studies were published during the last four years and 2023 was the year with the highest number of publications. Given that in the studies' search carried out on PubMed, Scopus, and Web of Science databases, no restrictions were made regarding the publication years, this result demonstrates that interest in the topic of this review is recent and is expanding.

In terms of geographical distribution three countries (i.e., China, India, and Saudi Arabia) contributed with more than two-thirds of the studies. Although the relative importance of Saudi Arabia is surprising, the contributions of China and India are in line with the results of other reviews related to the application of blockchain in different domains (e.g., [39, 40]).

Concerning the experimental setups, one study [16] did not collect experimental data. The remainder studies aimed to validate the proposed algorithms (i.e., six studies [21, 22, 24–26, 32]), to determine the performance of the proposed algorithms (i.e., thirteen studies [13–15, 17–20, 27, 29, 31, 34, 36, 37]), to compare the proposed algorithms with other algorithms described in scientific literature (i.e., four studies [23, 28, 30, 35]), and to compare the proposed federated learning algorithm with centralized learning algorithms (i.e., one study [33]).

Sixty percent of the studies designed the experimental setups considering the diagnosis of a specific pathology: (i) brain tumor prediction [28]; (ii) breast cancer detection [24, 26]; (iii) classification of synovial sarcoma cancer [17]; (iv) COVID-19 detection [14, 18, 22]; (v) lung cancer detection [15, 27]; (vi) lung cancer prediction [13, 21]; (vii) monkeypox detection [25]; (viii) osteosarcoma cancer detection [20]; (ix) pneumonia image classification [36]; and (x) skin lesion diagnosis [32].

Four studies (i.e., 16% of the studies) did not refer to any medical imaging modality, and two studies (i.e., 8% of the studies) aimed to address various medical imaging modalities: i) CT and X-ray, one study [34]; and (ii) MRI, PET, SPECT, one study [35]. In turn, the remainder 76% of the studies were focused on specific medical imaging modalities: (i) CT, ten studies [13–15, 18, 19, 21, 27, 29, 30, 33]; (ii) histopathology, four studies [17, 20, 24, 26]; (iii) X-ray, two studies [22, 36]; (iv) MRI, one study [28]; (v) skin images, one study [25]; and (vi) ultrasounds, one study [37].

On other hand, in terms of machine learning techniques, six techniques were used by more than one study: (i) CNN, eleven studies [13, 18, 20–22, 25, 28, 30, 34, 35, 37]; (ii) NN, three studies [29, 33, 36]; and (iii) ANN, two studies [23, 31]; (iv) CN, two studies [14, 17]; (v) DNN, two studies [15, 16]; and (vi) GRU, two studies [24, 26]. The remainder machine learning techniques being used were FCN [32], KELM [32], RCNN [15], and SVM [17].

In what concerns the datasets used to collect experimental data, only 72% of the studies considered publicly available medical imaging data sets [13, 19, 20, 23–28, 31, 34, 36], or local medical imaging datasets [14, 15, 18, 29, 33, 37]. The remainder studies either did not use medical images, but benchmarking images, or did not use any images, since did not perform any evaluation.

According to these results some included studies did not properly consider the specificities of the imaging modalities while others did not use adequate datasets for the algorithms' evaluations. Moreover, the translation of these algorithms in real-world conditions is not discussed as recommended by best practices for the reporting of machine learning algorithms applied to medical imaging [41]. These flaws impact the generalizability of the algorithms proposed by the studies.

Despite the limitations of this review (e.g., the dependency on the search and selection strategies) it is possible to conclude that the scientific literature does not provide evidence to translate for the clinical practice the use of both blockchain and machine learning techniques in medical imaging applications. This evidence must be provided by future research.

References

1. Cook, G.J.R., Goh, V.: What can artificial intelligence teach us about the molecular mechanisms underlying disease? Eur. J. Nucl. Med. Mol. Imaging **46**(13), 2715–2721 (2019)
2. Medeiros, E.P., Machado, M.R., de Freitas, E.D.G., da Silva, D.S., de Souza, R.W.R.: Applications of machine learning algorithms to support COVID-19 diagnosis using X-rays data information. Expert Syst. Appl. **238**(B), 122029 (2023)

3. Singh, S., Hoque, S., Zekry, A., Sowmya, A.: Radiological diagnosis of chronic liver disease and hepatocellular carcinoma: a review. J. Med. Syst. **47**(1), 73 (2023)

4. Sajed, S., Sanati, A., Garcia, J.E., Rostami, H., Keshavarz, A., Teixeira, A.: The effectiveness of deep learning vs. traditional methods for lung disease diagnosis using chest X-ray images: a systematic review. Appl. Soft Comput. **147**, 110817 (2023)

5. Narayan, V., Faiz, M., Mall, P.K., Srivastava, S.: A comprehensive review of various approach for medical image segmentation and disease prediction. Wireless Pers. Commun. **132**, 1819–1848 (2023)

6. Gupta, S.: Blockchain—The Foundation Behind Bitcoin. Wiley, New York (2017)

7. European Society of Radiology (ESR). ESR white paper: blockchain and medical imaging. Insights Imaging **12**(1), 82 (2021)

8. Aouedi, O., Sacco, A., Piamrat, K., Marchetto, G.: Handling privacy-sensitive medical data with federated learning: challenges and future directions. IEEE J. Biomed. Health Inform. **27**(2), 790–803 (2022)

9. Bashir, A.K., et al.: Federated learning for the healthcare metaverse: concepts, applications, challenges, and future directions. IEEE Internet Things J. **10**(24), 21873–21891 (2023)

10. Kumar, J., Singh, A.K.: Copyright protection of medical images: a view of the state-of-the-art research and current developments. Multimedia Tools Appl. **82**(28), 1–31 (2023)

11. Gomathi, L., Mishra, A.K., Tyagi, A.K.: Industry 5.0 for healthcare 5.0: opportunities, challenges and future research possibilities. In: 2023 7th International Conference on Trends in Electronics and Informatics (ICOEI), pp. 204–213. IEEE (2023)

12. Stephanie, V., Khalil, I., Atiquzzaman, M., Yi, X.: Trustworthy privacy-preserving hierarchical ensemble and federated learning in healthcare 4.0 with blockchain. IEEE Trans. Ind. Inf. **19**(7), 7936–7945 (2022)

13. Zerka, F., et al.: Blockchain for privacy preserving and trustworthy distributed machine learning in multicentric medical imaging (C-DistriM). IEEE Access **8**, 183939–183951 (2020)

14. Kumar, R., et al.: Blockchain-federated-learning and deep learning models for covid-19 detection using CT imaging. IEEE Sens. J. **21**(14), 16301–16314 (2021)

15. Kumar, R., et al.: An integration of blockchain and AI for secure data sharing and detection of CT images for the hospitals. Comput. Med. Imaging Graph. **87**, 101812 (2021)

16. Alamgeer, M., et al.: Privacy preserving image encryption with deep learning based IoT healthcare applications. Comput. Mater. Continua **2022**, 73(1), 1159–1175 (2022)

17. Arunachalam, P., et al.: Effective classification of synovial sarcoma cancer using structure features and support vectors. Comput. Mater. Continua **72**(2), 2521–2543 (2022)

18. Heidari, A., Toumaj, S., Navimipour, N.J., Unal, M.: A privacy-aware method for COVID-19 detection in chest CT images using lightweight deep conventional neural network and blockchain. Comput. Biol. Med. **145**, 105461 (2022)

19. Kumar, R., et al.: Blockchain and homomorphic encryption based privacy-preserving model aggregation for medical images. Comput. Med. Imaging Graph. **102**, 102139 (2022)

20. Nasir, M.U., Khan, S., Mehmood, S., Khan, M.A., Rahman, A.U., Hwang, S.O.: IoMT-based osteosarcoma cancer detection in histopathology images using transfer learning empowered with blockchain, fog computing, and edge computing. Sensors **22**(14), 5444 (2022)

21. Pawar, A.B., et al.: Implementation of blockchain technology using extended CNN for lung cancer prediction. Measur. Sens. **24**, 100530 (2022)

22. Ahmed, I., Chehri, A., Jeon, G.: Artificial Intelligence and Blockchain enabled smart healthcare system for monitoring and detection of COVID-19 in biomedical images. IEEE/ACM Trans. Comput. Biol. Bioinform. 1–10 (2023)

23. Albakri, A., Alqahtani, Y.M.: Internet of medical things with a Blockchain-assisted smart healthcare system using metaheuristics with a deep learning model. Appl. Sci. **13**(10), 6108 (2023)

24. Aldhyani, T.H., et al.: A secure internet of medical things framework for breast cancer detection in sustainable smart cities. Electronics **12**(4), 858 (2023)

25. Alruwaili, F.F., Alabduallah, B., Alqahtani, H., Salama, A.S., Mohammed, G.P., Alneil, A.A.: Blockchain enabled smart healthcare system using jellyfish search optimization with dual-pathway deep convolutional neural network. IEEE Access **11**, 87583–87591 (2023)

26. Chaudhury, S., Sau, K.: A blockchain-enabled internet of medical things system for breast cancer detection in healthcare. Healthcare Analytics **4**, 100221 (2023)

27. Heidari, A., Javaheri, D., Toumaj, S., Navimipour, N.J., Rezaei, M., Unal, M.: A new lung cancer detection method based on the chest CT images using Federated Learning and blockchain systems. Artif. Intell. Med. **141**, 102572 (2023)

28. Mohammad, F., Al Ahmadi, S., Al Muhtadi, J.: Blockchain-based deep CNN for brain tumor prediction using MRI scans. Diagnostics **13**(7), 1229 (2023)

29. Om Kumar, C.U., Gajendran, S., Balaji, V., Nhaveen, A., Sai Balakrishnan, S.: Securing health care data through blockchain enabled collaborative machine learning. Soft. Comput. **27**(14), 9941–9954 (2023)

30. Qamar, S.: Machine learning in cloud-based trust modeling in M-health application using classification with image encryption. Soft Comput. (2023)

31. Rahal, H.R., Slatnia, S., Kazar, O., Barka, E., Harous, S.: Blockchain-based multi-diagnosis deep learning application for various diseases classification. Int. J. Inf. Secur. **23**(1), 15–30 (2023)

32. Rajeshkumar, K., Ananth, C., Mohananthini, N.: Optimal hybrid image encryption with machine learning model for blockchain-assisted secure skin lesion diagnosis. Int. J. Eng. Trends Technol. **71**(6), 96–106 (2023)

33. Sai, S., Hassija, V., Chamola, V., Guizani, M.: Federated learning and NFT-based privacy-preserving medical data sharing scheme for intelligent diagnosis in smart healthcare. IEEE Internet Things J. **11**(4), 5568–5577 (2023)

34. Salim, M.M., Park, J.H.: Federated learning-based secure electronic health record sharing scheme in medical informatics. IEEE J. Biomed. Health Inform. **27**(2), 617–624 (2022)

35. Xiang, T., Zeng, H., Chen, B., Guo, S.: BMIF: privacy-preserving blockchain-based medical image fusion. ACM Trans. Multimed. Comput. Commun. Appl. **19**(1s), 1–23 (2023)

36. Yang, Y., Wei, J., Yu, Z., Zhang, R.A.: Trustworthy neural architecture search framework for pneumonia image classification utilizing blockchain technology. J. Supercomputing **80**(2), 1694–1727 (2023)

37. Guan, Y., Wen, P., Li, J., Zhang, J., Xie, X.: Deep learning blockchain integration framework for Ureteropelvic junction obstruction diagnosis using ultrasound images. Tsinghua Sci. Technol. **29**(1), 1–12 (2024)

38. McMahan, B., Moore, E., Ramage, D., Hampson, S., y Arcas, B.A.: Communication-efficient learning of deep networks from decentralized data. In: Artificial Intelligence and Statistics, pp. 1273–1282 (2017)

39. Joshi, P., Tewari, V., Kumar, S., Singh, A.: Blockchain technology for sustainable development: a systematic literature review. J. Glob. Oper. Strateg. Sourcing **16**(3), 683–771 (2023)

40. Jin, S., Chang, H.: The trends of blockchain in environmental management research: a bibliometric analysis. Environ. Sci. Pollut. Res. **30**(34), 81707–81724 (2023)

41. Cerdá-Alberich, L., et al.: MAIC–10 brief quality checklist for publications using artificial intelligence and medical images. Insights Imaging **14**(1), 11 (2023)

The Role of Electronic Health Records to Identify Risk Factors for Developing Long COVID: A Scoping Review

Ema Santos[1], Afonso Fernandes[1], Manuel Graça[1], and Nelson Pacheco Rocha[2]([✉]) ⓘ

[1] Department of Medical Sciences, University of Aveiro, Aveiro, Portugal
{emasofiabs,afonso.f,manuelsantorograca}@ua.pt
[2] IEETA, Department of Medical Sciences, University of Aveiro, Aveiro, Portugal
npr@ua.pt

Abstract. Coronavirus disease 2019 (COVID-19) was considered a global pandemic from December 2019 to May 2023. A subset of COVID-19 patients develops long-lasting sequelae, commonly referred to as long COVID. This scoping review aimed to identify risk factors for long COVID reported in multiple studies and to determine the role of the secondary use of Electronic Health Records to identify these risk factors. An electronic search was conducted on Scopus, PubMed, and Web of Science, and 46 studies were included in this review after the selection process. Thirty-one risk factors were identified, with the most referred ones being female sex, age, severity of infection and obesity. In terms of data collection, Electronic Health Records were used by 63.0% of the studies, although only 21.7% were retrospective studies exclusively based on the secondary use of Electronic Health Records data. These results show that the potential of clinical research based on the secondary use of data collected from Electronic Health Records is not yet fully achieved, despite the respective advantages when compared with other data collection methods such as remote surveys.

Keywords: Long COVID · Post-COVID-19 sequelae · Risk Factors · Electronic Health Records

1 Introduction

The global coronavirus 2019 (COVID-19) pandemic, caused by the severe acute respiratory syndrome coronavirus 2 (SARS-CoV-2), spread through the world at an unfathomable pace. It was considered a global pandemic from December 2019 to May 2023, when the World Health Organization (WHO) declared the end of COVID-19 as a global health emergency.

A subset of COVID-19 patients developed long-lasting sequelae, commonly referred to as post-COVID-19 sequelae or long COVID. Long COVID is commonly used when referring to the ongoing symptomatic COVID-19 (from 4 to 12 weeks) and can be defined as "signs and symptoms that develop during or after an infection consistent with COVID-19, continue for more than 12 weeks and are not explained by an alternative

Á. Rocha et al. (Eds.): WorldCIST 2024, LNNS 986, pp. 118–128, 2024.
https://doi.org/10.1007/978-3-031-60218-4_12

diagnosis" [1]. It is estimated that 10% of infections lead to long COVID, although this number is likely to be much higher due to many undocumented cases. And as diagnostic and treatment options are currently insufficient, a significant proportion of patients may have lifelong disabilities if no action is taken [2].

Pre-existing comorbidities, such as type 2 diabetes, respiratory diseases, and mental disorders are frequently mentioned as risk factors of long COVID in current literature, along with female sex and old age [2, 3]. The first objective of the present scoping review was to systematize risk factors of long COVID that have been identified by clinical studies.

Electronic Health Records systems aim to electronically collect person-based clinical information derived from administrative, demographic, and clinical processes which are needed to diagnose and manage individual's illnesses [4]. Consequently, these systems yield a rich source of clinical information which can be used to support the planning and development of healthcare services and to monitor their quality, as well as to facilitate clinical research [4]. The secondary use of clinical information holds enormous potential for clinical research, namely, to improve the detection and surveillance of emerging diseases, or to provide new insights into causes and consequences of existing diseases [5]. Therefore, this scoping review's second objective, which is absent from current scientific literature, was to determine the role of the secondary use of Electronic Health Records in identifying risk factors of long COVID.

2 Methods

The research for this scoping review relied on retrieving articles from three online databases (i.e., Scopus, Web of Science and PubMed) conform with the following search query: "longcovid" or "long covid" or "post covid19" or "long term covid19".

The retrieved articles were filtered through a series of inclusion and exclusion criteria. The authors aimed to include peer reviewed articles reporting on the identification of long COVID risk factors. In turn, the authors excluded articles not published in English, Portuguese or Spanish, articles that did not report primary studies (i.e., reviews, surveys, overviews, or editorials), articles without information regarding authors or abstracts, and articles whose full texts were not available.

The included articles were analysed, and syntheses were prepared to systematise (i) the geographical localization of the studies, (ii) the characteristics of the studies' design, including an analysis of how Electronic Health Records were used to support data collection, and (iii) the risk factors for developing long COVID.

3 Results

3.1 Studies' Selection

The access date was March 3rd, 2023. As a result, 3398 articles were obtained (i.e., 1671 from Scopus, 1395 from Web of Science, and 332 from PubMed).

During the first phase of the screening procedure, 1541 references were excluded (i.e., 1058 duplicates, 193 reviews, surveys, overviews, and editorials, and 290 references lacking abstracts or author's names). Then, after analysing titles and abstracts, 1794 references were removed for not complying with the eligibility criteria. Finally, the authors assessed the full texts of the remainder 63 references of which 17 were excluded according to the eligibility criteria. Therefore, 46 studies [6–51] were included in this scoping review.

3.2 Geographical Distribution

Table 1 presents the geographical characteristics of the included studies. Most of the studies (i.e., 42 studies) were national studies, while only four studies [10, 20, 44, 46] were conducted at international level. From these international studies, one of them [46] considered 28 countries (i.e., Switzerland, Israel, and all the Member States of the European Union except Ireland).

Table 1. Geographical distribution of the included studies.

National Studies	
Countries with one study	Brazil [48], China [30], Croatia [13], Denmark [49], Ecuador [27], Egypt [7], France [29], Germany [25], India [12], Malta [15], Morocco [21], Nigeria [8], Sweden [26], Turkey [14]
Countries with two studies	Russia [33, 34]
Countries with three studies	Israel [31, 45, 50], Poland [17, 18, 36], Spain [6, 22, 23], UK [42, 43, 47]
Countries with more than three studies	Italy, nine studies [9, 11, 16, 19, 32, 37–40]; USA, five studies [24, 28, 35, 41, 51]
International studies	
UK, USA, and Sweden [10]; Israel, Italy, Spain and Switzerland [20]; UK and USA [44]; 28 countries [46]	

3.3 Studies' Design

The included studies followed different definitions of long COVID and considered different time frames to include participants (e.g., two weeks, three months, six months, or longer timeframes). Moreover, the sample sizes of the included studies (Table 2) were quite heterogeneous (i.e., varied from 106 participants [26] to 2,430,729 participants [42]) and more than half of the studies included less than one thousand participants.

Seventeen articles [8–13, 15, 16, 18, 21, 27, 32, 35, 37, 39, 44, 46] did not mention the use of data from Electronic Health Records. The research data of the studies reported by these articles were collected by: (i) clinical assessments [8, 9, 11, 18, 32, 39]; (ii) face-to-face questionnaires [13, 16, 37]; (iii) questionnaires delivered by special purpose

Table 2. Sample size for each included article.

Sample size	Number	Studies
[1–100]	0	
[101–1000]	25	[6–9, 12–14, 16, 18, 19, 21, 24–26, 28, 29, 31–33, 37–41, 48]
[0–10]	14	[10, 11, 15, 17, 20, 22, 23, 27, 30, 34, 36, 43, 45, 47]
[0–100]	3	[35, 44, 46]
[100, 001–1, 000, 000]	3	[49–51]
>1,000,000	1	[42]

apps [10, 35]; (iv) online questionnaires [15, 27, 44, 46]; (v) questionnaires delivered by electronic mail [21]; and (vi) telephone call questionnaires [12].

Furthermore, in 19 studies [6, 14, 17, 19, 20, 22–24, 26, 28–31, 33, 34, 40, 41, 47, 48] abstracts of Electronic Health Records (e.g., demographics, clinical characteristics, symptoms on admission, disease severity, supportive therapies or even medical imaging studies of the participants) complemented data collected by other means, such as (i) clinical evaluations [17, 19, 20, 26, 31, 40], (ii) face-to-face interviews [29, 41, 47], (iii) telephone interviews [6, 22–24, 30, 33, 34, 48], (iv) face-to-face and telephone interviews [14], and (v) on-line questionnaires [28].

Finally, 10 articles [7, 25, 36, 38, 42, 43, 45, 49–51] reported retrospective studies exclusively supported on the secondary use of data available on Electronic Health Records. Three of these [7, 25, 38] were single centre studies supported on the Electronic Health Records of specific hospitals form El-Minia, Egypt [7], Dresden, Germany [25] and Rome, Italy [38]. The remainder seven studies [36, 42, 43, 45, 49–51] were retrospective studies supported by national Electronic Health Records, as presented in Table 3. These studies represented 94.7% of the participants, and four of them [42, 49–51] occupied the first four positions in terms of number of participants, immediately followed by three studies whose data were collected remotely (i.e., online questionnaires: 54,960 participants [44] and 49,044 participants [46]; or remote apps: 16,091 participants [35]).

Considering the national Electronic Health Records being used (Table 3), the STOP-COVID registry includes only patients from Poland and contains medical information of patients presenting to health centres for persistent clinical symptoms after COVID-19 and subsequent follow-up visits at 3 and 12 months [17]. In turn, the Clinical Practice Research Datalink (CPRD) Aurum and UK National Primary Care Electronic Health Record are related to the primary care provision in UK. The first is an anonymized database containing routinely collected data from primary care practices, and the second is a transversal Electronic Health Record covering more than 95% of the UK population [43]. Moreover, the Maccabi Healthcare Services Database is the centralized Electronic Health Record of the second largest healthcare organization in Israel, which covers more than two million citizens [45], while the Danish National Patient Registry provides data on all discharges from Danish hospitals. Finally, INSIGHT and OneFlorida + are two large clinical research networks. The INSIGHT covers a diverse patient population in the

Table 3. Retrospective studies based on national Electronic Health Records.

Article	Participants	Dataset
[36]	1,847	STOP-COVID Registry
[42]	2,430,729	Clinical Practice Research Datalink Aurum
[43]	6,907	UK National Primary Care Electronic Health Record
[45]	3,240	Maccabi Healthcare Services Database
[49]	49,044	Danish National Patient Registry
[50]	180,759	Maccabi Healthcare Services Database
[51]	100,450	INSIGHT and OneFlorida +

New York City Metropolitan Area and the OneFlorida + is a partnership of 14 academic institutions and health systems across Florida, Georgia, and Alabama with longitudinal patient-level Electronic Health Record data for more than 20 million patients [51].

3.4 Risk Factors

Risk factors that were identified by at least two studies are shown in Table 4, along with the number of studies that support them.

Table 4. Risk factors that were identified by at least two studies. Some risk factors have been grouped for evidence according to their close associations.

Risk Factor	Number	Studies
Female sex (and female gender)	22	[10, 13–15, 17, 20, 22, 24, 27–29, 35–37, 39–43, 47, 48, 50]
Severity of initial SARS-CoV-2 infection (including hospitalization and a higher number of initial symptoms)	20	[8, 10, 12, 13, 17, 18, 21, 23, 24, 26, 28–30, 36, 37, 40, 45–47, 50]
Increasing age (elders vs adults, and teenager's vs children)	17	[9, 10, 14, 16, 24, 27, 33–37, 39, 42, 43, 45, 46, 50]
Obesity, overweight, and sedentarism	12	[11, 14, 16, 18, 20, 25, 27, 36, 42, 43, 48, 50]
Mental disorders (including anxiety, depression, and psychiatric history)	8	[14, 28, 36, 39, 42, 44, 47, 50]

(continued)

Table 4. (*continued*)

Risk Factor	Number	Studies
Heart diseases (including hypertension, myocardial disease, and ischemic heart disease)	6	[7, 12, 29, 36, 38, 50]
Comorbidities (in general)	6	[14, 19, 20, 36, 39, 42]
Respiratory diseases (including chronic obstructive pulmonary disorder and asthma)	5	[7, 36, 41–43]
Socioeconomic deprivation, neighbourhood deprivation, lower education levels	5	[35, 42, 46, 47, 51]
Diabetes type 2	4	[7, 12, 41, 49]
Ethnic minorities	3	[35, 42, 43]
Smoking or former smoking	3	[36, 42, 50]
Allergies	2	[16, 33]
Sleep Disorders (including insomnia, poor sleep quality and working night shifts)	2	[31, 36]

Moreover, the following risk factors were identified by single studies: air toxicants [51], arthralgia [18], asthenia [39], celiac disease [42], benign prostatic hyperplasia [42], dementia [24], dehydration [26], Epstein-Barr virus viremia [41], erectile dysfunction [42], fibromyalgia [42], hypercholesterolemia [48], learning disability [42], migraine and headache disorder [42], multiple sclerosis [42], original SARS-CoV-2 variant (compared to Omicron) [30], SARS-CoV-2 RNAemia [41], and vaccination against SARS-CoV-2 [27].

4 Discussion

The studies analysed in this review included participants from many different countries. In terms of the number of studies, Italy stands out as the most referenced, followed by the USA. At the start of the pandemic in early 2020, Italy had the highest number of deaths by COVID-19, and was a global COVID-19 hotspot, being one of the top ten countries to suffer the most from the first months of the pandemic.

In total, 31 risk factors were identified by the included studies. When considering the risk factors reported by the highest number of studies, twelve major risk factors for developing long COVID were identified by at least four studies (Fig. 1). Female sex, increasing age, severity of initial SARS-CoV-2 infection and obesity were reported by the highest number of articles. Many studies attempt to explain why females can develop post-COVID symptoms to a greater extent than males. Some biological differences, such as the expression of angiotensin-converting enzyme-2 (ACE2) – the viral receptor for SARS-CoV-2 –, and the production of pro-inflammatory cytokines after viral infection, are hypothesized to play a role in higher development of symptoms. Additionally, the higher prevalence of pain syndromes in females, along with other factors like higher psychological stress, higher depressive levels, and poorer sleep quality, have also been associated with some post-COVID symptoms in females [22, 39]. It is possible then that female sex as a risk factor is acting together with other long COVID risk factors more predominant in females, making it the most impactful risk factor found.

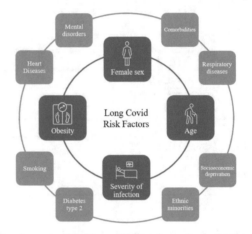

Fig. 1. Major risk factors for developing long COVID.

Regarding age as a risk factor, increasing age is associated with both COVID-19 disease severity and higher risk of post-COVID-19 conditions. This may be explained by the natural aging of the immune system, and the consequent diminished immune functions. Furthermore, age-related comorbidities are associated with a higher risk of developing long COVID. Besides adults, older children are also more prone to develop sequelae than younger children. It's possible that they have been more negatively impacted by the pandemic. On the other hand, some symptoms are harder to detect in younger children [16]. As for obesity, it is highly related to both a more severe course of the acute phase and a higher risk of developing long COVID. This is partly due to the systemic inflammatory state of obesity, as well as endothelial dysfunction and increased expression of ACE2 that are associated with overweight individuals [16].

Other relevant risk factors include mental disorders (such as anxiety and depression), pre-existing comorbidities (such as heart diseases, respiratory diseases, and diabetes type 2), and smoking. There were also studies focused on the impact of COVID-19 on a social level, reporting that individuals belonging to ethnic minorities and those living in socioeconomic deprivation are at higher risk of developing long COVID. The remaining risk factors present low relative evidence, which doesn't mean they should be disregarded. Ultimately, many of the listed risk factors relate to each other and contribute jointly to long COVID risk.

The number of participants varied greatly from study to study, with the minimum sample size being 106 participants [19] and the maximum being 2,430,729 participants [42]. Moreover, different methods of data collection were identified, including Electronic Health Records, clinical assessments, face-to-face questionnaires, questionnaires delivered by special purpose apps, online questionnaires, questionnaires delivered by electronic mail, and telephone call questionnaires.

In what concerns the use of data from Electronic Health Records, 37.0% of the studies did not mention their use, 41.3% extracted data from Electronic Health Records to complement data collected by other methods, including remote questionnaires (e.g., online questionnaires), and 21.7% were retrospective studies exclusively supported on the secondary use of data available on Electronic Health Records. In these retrospective studies, which represented 94.7% of the participants, it was possible to identify seven national Electronic Health Records: Clinical Practice Research Datalink Aurum and UK National Primary Care Electronic Health Record (UK), INSIGHT and OneFlorida + (USA), Danish National Patient Registry (Denmark), Maccabi Healthcare Services Database (Israeli), and STOP-COVID Registry (Poland).

This scoping review has several limitations related to the analysis of the articles included that must be considered when analysing the results. It has only been over three years since the start of the COVID-19 pandemic, and studies on long COVID patients are limited by the time since initial infection. It will only be possible to understand the full extent of the post-COVID-19 sequelae once there is clinical information concerning the entire natural lifespan of long COVID patients. On the same note, not all the studies followed the same definition of post-COVID-19 or included patients within the same timeframe, meaning this scoping review joins long COVID patients at different intervals, such as two weeks, three months, six months and longer. Moreover, a significant number of studies were based on self-diagnosed symptoms, which can be entirely subjective depending on the patient's knowledge and understanding of the situation. The few studies that focused on clinically diagnosed manifestations did not usually follow broad objectives, instead choosing only one symptom to investigate. Additionally, there may be an overlap in symptoms, particularly due to comorbidities, which further complicates correct diagnoses.

5 Conclusion

The present scoping review managed to find many risk factors for developing long COVID and determined the following as the most evidenced risk factors: female sex, increasing age, obesity, and severity of initial SARS-CoV-2 infection. Additional risk factors reported in multiple studies with abundant evidence include pre-existing comorbidities, such as respiratory diseases, heart diseases and diabetes type 2, mental disorders, smoking, ethnic minorities, and socioeconomic deprivation.

According to the results, only 21.7% were retrospective studies exclusively supported on the secondary use Electronic Health Records data, although the richness and correctness of the data collected from Electronic Health Records are necessarily higher than the data collected by other methods to include large numbers of participants such as remote surveys using special purpose digital applications or online questionnaires. This means that the full potential of the secondary use of data from Electronic Health Records is not yet achieved.

Acknowledgments. This study was carried out within the scope of the course unit Clinical Information Management of the Master's in Clinical Bioinformatics at the University of Aveiro.

References

1. Shah, W., Hillman, T., Playford, E.D., Hishmeh, L.: Managing the long term effects of covid-19: summary of NICE, SIGN, and RCGP rapid guideline. BMJ **372** (2021)
2. Davis, H.E., McCorkell, L., Vogel, J.M., Topol, E.J.: Long COVID: major findings, mechanisms and recommendations. Nat. Rev. Microbiol. **21**, 133–146 (2023)
3. Koc, H.C., Xiao, J., Liu, W., Li, Y., Chen, G.: Long COVID and its Management. Int. J. Biol. Sci. **18**(12), 4768–4780 (2022)
4. Williams, J.G.: The use of clinical information to help develop new services in a district general hospital. Int. J. Med. Informatics **56**(1–3), 151–159 (1999)
5. Näher, A.F., et al.: Secondary data for global health digitalisation. Lancet Digital Health, **5**(2), e93-e101 (2023)
6. Fernández-de-Las-Peñas, C., et al.: Diabetes and the risk of long-term post-COVID symptoms. Diabetes **70**(12), 2917–2921 (2021)
7. Mady, A.F., Abdelfattah, R.A., Kamel, F.M., Abdel Naiem, A.S.M., AbdelGhany, W.M., Abdelaziz, A.O.: Predictors of long covid 19 syndrome. Egyptian J. Hospital Med. **85**(2), 3604–3608 (2021)
8. Osikomaiya, B., et al.: Long COVID': persistent COVID-19 symptoms in survivors managed in Lagos State, Nigeria. BMC Infect. Diseases 21(1), 1–7 (2021)
9. Peghin, M., et al.: Post-COVID-19 symptoms 6 months after acute infection among hospitalized and non-hospitalized patients. Clinical Microbiol. Infect. **27**(10), 1507–1513 (2021)
10. Sudre, C.H., et al.: Attributes and predictors of long COVID. Nat. Med. **27**(4), 626–631 (2021)
11. Vimercati, L., et al.: Association between long COVID and overweight/obesity. J. Clin. Med. **10**(18), 4143 (2021)
12. Arjun, M.C., et al.: Characteristics and predictors of Long COVID among diagnosed cases of COVID-19. Plos one **17**(12), e0278825 (2022)
13. Banić, M.,et al.: Risk factors and severity of functional impairment in long COVID: a single-center experience in Croatia. Croatian Med. J. **63**(1), 27–35 (2022)
14. Baris, S.A., et alk.: The predictors of long–COVID in the cohort of Turkish Thoracic Society–TURCOVID multicenter registry: one year follow–up results. Asian Pacific J. Tropical Med. **15**(9), 400–409 (2022)
15. Baruch, J., Zahra, C., Cardona, T., Melillo, T.: National long COVID impact and risk factors. Public Health **213**, 177–180 (2022)
16. Buonsenso, D., et al.: The prevalence, characteristics and risk factors of persistent symptoms in non-hospitalized and hospitalized children with sars-cov-2 infection followed-up for up to 12 months: a prospective, cohort study in Rome, Italy. J. Clin. Med. **11**(22), 6772 (2022)
17. Chudzik, M., Babicki, M., Kapusta, J., Kałuzinska-Kołat, Z., Kołat, D., Jankowski, P.: Long-COVID Clinical Features and Risk Factors: a Retrospective Analysis of Patients from the STOP-COVID Registry of the PoLoCOV Study. Viruses **14**(8), 1755 (2022)

18. Chudzik, M., Lewek, J., Kapusta, J., Banach, M., Jankowski, P., Bielecka-Dabrowa, A.: Predictors of long COVID in patients without comorbidities: data from the polish long-covid cardiovascular (polocov-cvd) study. J. Clin. Med. **11**(17), 4980 (2022)
19. Cristillo, V., Pilotto, A., Cotti Piccinelli, S., Bonzi, G., et al.: Premorbid vulnerability and disease severity impact on Long-COVID cognitive impairment. Aging Clin. Experim. Res. **34**, 257–260 (2022)
20. Daitch, V., Yet al.: Characteristics of long-COVID among older adults: A cross-sectional study. Inter. J. Infect. Diseases **125**, 287–293 (2022)
21. El Otmani, H., Nabili, S., Berrada, M., Bellakhdar, S., El Moutawakil, B., Abdoh Rafai, M.: Prevalence, characteristics and risk factors in a Moroccan cohort of Long-Covid-19. Neurol. Sci. **43**(9), 5175–5180
22. Fernández-de-Las-Peñas, C., et al.: Female sex is a risk factor associated with long-term post-COVID related-symptoms but not with COVID-19 symptoms: the LONG-COVID-EXP-CM multicenter study. J. Clin. Med. **11**(2), 413 (2022)
23. Fernández-de-Las-Peñas, C., et al.: Symptoms experienced at the acute phase of SARS-CoV-2 infection as risk factor of long-term post-COVID symptoms: the LONG-COVID-EXP-CM multicenter study. Int. J. Infect. Dis. **116**, 241–244 (2022)
24. Frontera, J.A., et al.: Life stressors significantly impact long-term outcomes and post-acute symptoms 12-months after COVID-19 hospitalization. J. Neurol. Sci. **443**, 120487 (2022)
25. Heubner, L., et al.: Extreme obesity is a strong predictor for in-hospital mortality and the prevalence of long-COVID in severe COVID-19 patients with acute respiratory distress syndrome. Sci. Rep.**12**(1), 18418
26. Hultström, M., et al.: Dehydration is associated with production of organic osmolytes and predicts physical long-term symptoms after COVID-19: a multicenter cohort study. Critical Care **26**(1), 1–9 (2022)
27. Izquierdo-Condoy, J.S., et al.: Long COVID at different altitudes: a countrywide epidemiological analysis. Inter. J. Environ. Res. Public Health **19**(22), 14673 (2022)
28. Knight, D.R., Munipalli, B., Logvinov, I.I., Halkar, M.G., Mitri, G., Hines, S.L.: Perception, prevalence, and prediction of severe infection and post-acute sequelae of COVID-19. Am. J. Med. Sci. **363**(4), 295–304 (2022)
29. Ko, A.C.S.,et al.: Number of initial symptoms is more related to long COVID-19 than acute severity of infection: A prospective cohort of hospitalized patients. Inter. J. Infect. Diseases **118**, 220–223 (2022)
30. Liao, X., et al.: Long-term sequelae of different COVID-19 variants: the original strain versus the Omicron variant. Global Health Med. **4**(6), 322–326 (2022)
31. Margalit, I., Yelin, D., Sagi, M., Rahat, M. M., Sheena, L., Mizrahi, N., ... Yahav, D.: Risk factors and multidimensional assessment of long coronavirus disease fatigue: a nested case-control study. Clinical Infectious Diseases, 75(10), 1688–1697 (2022)
32. Mazza, M.G., Palladini, M., Villa, G., De Lorenzo, R., Querini, P.R., Benedetti, F.: Prevalence, trajectory over time, and risk factor of post-COVID-19 fatigue. J. Psychiatr. Res.Psychiatr. Res. **155**, 112–119 (2022)
33. Osmanov, I.M., et al.: Risk factors for post-COVID-19 condition in previously hospitalised children using the ISARIC Global follow-up protocol: a prospective cohort study. European Resp. J. **59**(2) (2022)
34. Pazukhina, E., et al.: Prevalence and risk factors of post-COVID-19 condition in adults and children at 6 and 12 months after hospital discharge: a prospective, cohort study in Moscow (StopCOVID). BMC Med. **20**(1), 244 (2022)
35. Perlis, R.H., et al.: Prevalence and correlates of long COVID symptoms among US adults. JAMA Netw. Open **5**(10), e2238804-e2238804 (2022)

36. Pływaczewska-Jakubowska, M., Chudzik, M., Babicki, M., Kapusta, J., Jankowski, P.: Lifestyle, course of COVID-19, and risk of Long-COVID in non-hospitalized patients. Front. Med. **9**, 1036556 (2022)
37. Righi, E., et al.: Determinants of persistence of symptoms and impact on physical and mental wellbeing in Long COVID: a prospective cohort study. J. Infect. **84**(4), 566–572 (2022)
38. Rinaldi, R., Basile, M., Salzillo, C., Grieco, D. L., Caffè, A., Masciocchi, C.: Gemelli against COVID Group.: Myocardial injury portends a higher risk of mortality and long-term cardiovascular sequelae after hospital discharge in COVID-19 survivors. J. Clin. Med. **11**(19), 5964 (2022)
39. Sansone, D., et al.: Persistence of symptoms 15 months since COVID-19 diagnosis: prevalence, risk factors and residual work ability. Life **13**(1), 97 (2022)
40. Spinicci, M., et al.: Infection with SARS-CoV-2 variants is associated with different long COVID phenotypes. Viruses **14**(11), 2367 (2022)
41. Su, Y., et al.: Multiple early factors anticipate post-acute COVID-19 sequelae. Cell **185**(5), 881–895 (2022)
42. Subramanian, A., et al.: Symptoms and risk factors for long COVID in non-hospitalized adults. Nat. Med. **28**(8), 1706–1714 (2022)
43. Thompson, E.J., Williams, D.M., Walker, A.J., Mitchell, R.E., Niedzwiedz, C.L., Yang, T.C.: Long COVID burden and risk factors in 10 UK longitudinal studies and electronic health records. Nat. Commun. **13**(1), 3528 (2022)
44. Wang, S., et al.: Associations of depression, anxiety, worry, perceived stress, and loneliness prior to infection with risk of post–COVID-19 conditions. JAMA Psych. **79**(11), 1081–1091 (2022)
45. Adler, L., et al.: Long COVID symptoms in Israeli children with and without a history of SARS-CoV-2 infection: a cross-sectional study. BMJ Open **13**(2), e064155 (2023)
46. Bovil, T., Wester, C.T., Scheel-Hincke, L.L., Andersen-Ranberg, K.: Risk factors of post-COVID-19 conditions attributed to COVID-19 disease in people aged≥ 50 years in Europe and Israel. Public Health **214**, 69–72 (2023)
47. Daines, L., et al.: Characteristics and risk factors for post-COVID-19 breathlessness after hospitalisation for COVID-19. ERJ Open Res. **9**(1) (2023)
48. Lapa, J., Rosa, D., Mendes, J.P.L., Deusdará, R., Romero, G.A.S.: Prevalence and associated factors of post-COVID-19 syndrome in a Brazilian cohort after 3 and 6 months of hospital discharge. Int. J. Environ. Res. Public Health **20**(1), 848 (2023)
49. Nørgård, B.M., Zegers, F.D., Juhl, C.B., Kjeldsen, J., Nielsen, J.: Diabetes mellitus and the risk of post-acute COVID-19 hospitalizations—a nationwide cohort study. Diabet. Med. **40**(2), e14986 (2023)
50. Tene, L., Bergroth, T., Eisenberg, A., David, S.S.B., Chodick, G.: Risk factors, health outcomes, healthcare services utilization, and direct medical costs of patients with long COVID. Int. J. Infect. Dis. **128**, 3–10 (2023)
51. Zhang, Y., et al.: Identifying environmental risk factors for post-acute sequelae of SARS-CoV-2 infection: An EHR-based cohort study from the recover program. Environ. Adv. **11**, 100352 (2023)

Machine Learning Approaches to Support Medical Imaging Diagnosis of Pancreatic Cancer – A Scoping Review

Florbela Tavares[1], Gilberto Rosa[1], Inês Henriques[1], and Nelson Pacheco Rocha[2](✉) (iD)

[1] Department of Medical Sciences, University of Aveiro, Aveiro, Portugal
{florbelatavares,gilberto.rosa,ines.henriques}@ua.pt
[2] IEETA, Department of Medical Sciences, University of Aveiro, Aveiro, Portugal
npr@ua.pt

Abstract. The early diagnosis of pancreatic cancer through medical imaging modalities could be an important breakthrough in the increasing of the survival rate. The purpose of this scoping review was to systematize the application of machine learning models to assist the medical imaging diagnosis of pancreatic cancer. An electronic search was conducted on PubMed, Scopus, Web of Science and Association for Computing Machinery, and 20 studies were included in this review after the selection process. Eleven different machine models were identified in the included studies, and, among these, convolutional neural network (CNN) the most referred (i.e., six studies). In general, the included studies present high values in terms of accuracy, sensitivity, and specificity of the machine learning algorithms. However, the included studies only considered retrospective data. This means that randomized clinical trials are required to translate machine learning implementations to clinical practice.

Keywords: Pancreatic Cancer · Machine Learning · Diagnosis · Medical Imaging · Scoping Review

1 Introduction

Pancreatic cancer has high incidence rates [1] and represents about 3% of all cancer diagnoses and accounts for 5% of cancer related deaths [2], with pancreatic ductal adenocarcinoma representing more than 90% of all pancreatic cancers [2]. Early diagnosis of pancreatic cancer is difficult with only about 10% of diagnosis made at an early stage [2]. The lateness in diagnosis results from the fact that most patients do not present symptoms at an early stage of the disease, and when diagnosis is made, the disease has already progressed into a metastatic stage [2, 3]. The five-year survival rate for pancreatic cancer is 13% [4], but when the tumor is less than 1cm in diameter, the five-year survival rate might increase to 80% [5]. Therefore, early diagnosis of pancreatic cancer can be crucial for a patient's prognosis.

Á. Rocha et al. (Eds.): WorldCIST 2024, LNNS 986, pp. 129–138, 2024.
https://doi.org/10.1007/978-3-031-60218-4_13

In general, the main diagnosis for pancreatic cancer relies on medical imaging modalities such as endoscopic ultrasound (EUS), magnetic resonance imaging (MRI), computed tomography (CT) and positron emission tomography (PET) [2]. CT scanning is the most used medical imaging tool for pancreatic cancer diagnosis. CT can also predict if the tumor is unresectable [2] and, since it takes interval images, allowing for a 3D reconstruction, it can also be useful for pre-operative planning. In turn, EUS allows pancreas observation with a high spatial resolution, having better diagnosis performance than CT, however it is highly dependent on the clinicians' experience. Furthermore, this is an invasive procedure where a gastrointestinal EUS involves patient sedation [2]. PET alone doesn't provide any advantage for pancreatic cancer diagnosis than other methods, but when combined with CT is able to detect metabolism information that can be a powerful tool to support diagnoses. Nonetheless, studies suggest that PET/CT performs equal as CT alone [2].

The success of machine learning (ML) models in numerous pattern recognition applications [6] has created optimism and high expectations for significant developments in health care. The potential for using ML with medical imaging analysis to give support to clinicians by enhancing the accuracy and efficiency of various diagnostic and treatment processes has promoted a general optimism in developing and using computer-aided diagnosis in clinical practice [7]. In this respect, this scoping review aimed to provide the reader with the state of the art of the different ML models that might support the medical imaging diagnosis of pancreatic cancer. With the difficulty posed for pancreatic cancer diagnosis and the fact that diagnosis can be highly dependent on the clinicians' expertise, this review aims to understand how ML models can be useful to improve pancreatic cancer diagnosis.

2   Methods

The scoping review was supported on a protocol that defined the methods to be used and the steps to be taken, namely, (i) databases search; (ii) inclusion and exclusion criteria; (iii) selection procedures; and (iv) synthesis and reporting.

2.1   Databases Search

Four different online databases were considered to retrieve the studies to be included in the review: (i) PubMed; (ii) Scopus; (iii) Web of Science (WoS); and (iv) Association for Computing Machinery (ACM). Table 1 presents the research queries that were used for the different databases.

Table 1. Research queries used for PubMed, Scopus, WoS and ACM.

Database	Search Query
PubMed	(Artificial Intelligence) OR (Machine learning) AND (pancreas) AND (cancer) AND (diagnostic)
Scopus	TITLE-ABS-KEY ("artificial intelligence" OR "machine learning" AND pancreas AND cancer AND diagnostic)
WoS	"Artificial intelligence" OR "Machine learning") AND pancreas AND cancer AND diagnostic
ACM	[All: artificial intelligence] AND [All: cancer] AND [All: pancreas] AND [All: diagnostic]

2.2 Inclusion and Exclusion Criteria

The most relevant criterium for studies' inclusion was if their references mentioned either ML or Artificial Intelligence (AI) models, aiming to support pancreatic cancer diagnosis, specifically diagnosis made through medical imaging modalities.

As for the exclusion criteria, articles not written in English, articles without their authors' information or without abstracts or articles reporting secondary studies such as systematic or scoping reviews were excluded, as well as editorial articles, conference papers and book chapters. Furthermore, studies focusing differential diagnosis between pancreatic diseases were also excluded.

Table 2 shows both the inclusion and exclusion criteria.

Table 2. Inclusion and exclusion criteria.

Criteria	Specification of Criteria
Inclusion	Articles focused on the application of ML or AI models to support pancreatic cancer diagnosis based on medical imaging
Exclusion	Articles not written in English Articles without authors or abstracts Articles reporting secondary studies Editorial articles, book chapters and conference papers Articles reporting studies focused on differential diagnosis between pancreatic diseases

2.3 Selection Procedures

In this step, considering the eligibility criteria established, the selection of studies was conducted. This process consisted of three main steps: (i) identification, where the articles were extracted from the selected databases; (ii) screening, by the analysis of the titles and abstract, the articles not conformed with the inclusion criteria were excluded; (iii) eligibility, where the full text was analyzed according to the inclusion and exclusion criteria.

2.4 Synthesis and Reporting

Based on the demographic data, the authors prepared a synthesis of the characteristics of the included studies, considering (i) the distribution of the studies by publication years and (ii) the geographical distribution of the studies, considering the first author affiliation.

Moreover, a tabular synthesis was prepared to systematize the experimental characteristics of the studies, including (i) medical imaging modalities, (ii) ML models, (iii) population of the studies, and (iv) performance measures. The performance measures analyzed were accuracy, sensitivity, and specificity. Accuracy provides an overall assessment of the model's correctness, while sensitivity and specificity delve into the model's ability to correctly identify positive and negative instances, respectively [8, 9].

3 Results

3.1 Selection Process

The search on the selected databases was conducted in March 2023 and identified 582 references: (i) 253 from PubMed; (ii) 279 from Scopus; (iii) 44 from WoS; and (iv) six from ACM. After the selection process (Table 3) 20 studies were considered for this review [10–29].

Table 3. Selection process.

Excluded	
	102 duplicates
	19 not published in English
	117 secondary studies (reviews or surveys)
	29 conference papers
	1 book chapter
	284 excluded by the title and abstract analysis
	10 excluded by the full text analysis
Included	
	20 studies

3.2 Demographic Characteristics of the Included Studies

The included studies were published between 2008 (i.e., one study [10]) and the first two months of 2023 (i.e., two studies [28, 29]). Six studies [21–27] were published in 2022, the year with the highest number of publications.

This systematic review included twenty studies with a widespread throughout the world, involving studies from America, Europe, and Asia, with the USA and China being in the lead (Fig. 1).

3.3 Experimental Characteristics of the Included Studies

Table 4 shows the summary of each study included in this review, namely the medical imaging modalities and the ML models, as well as the population included in the studies and the respective results in terms of accuracy, sensitivity, and specificity.

Eleven different ML models were identified: deep learning (DL), namely convolutional neural network (CNN) and faster region-convolutional neural network (Faster R-CNN), extreme gradient boosting (XGBoost), end-to-end deep learning (E2EDL), a hybrid model combining support vector machine and random forest (H-SVM-RF), k-nearest neighbour (KNN), multilayer perceptron (MLP), Naïve Bayes (NB), random forest (RF), and support vector machines (SVM).

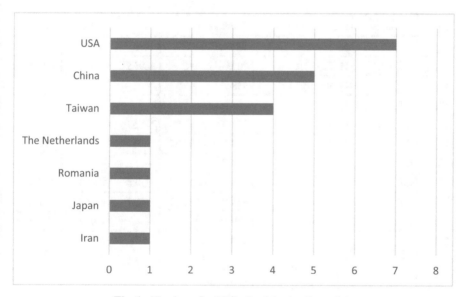

Fig. 1. Number of published articles by Countries.

Considering the population included in the studies, there was a great heterogeneity and varied from 40 patients and 40 healthy controls [11] to 1217 patients and 1593 healthy controls [28, 29].

Four articles [12, 13, 16, 22] did not report performance values. The remainder of the articles reported the achieved values of accuracy or sensitivity and specificity. In general, the reported values are high. However, only retrospective patients' data were analyzed.

4 Discussion

This scoping review identified 20 studies focused on the development of ML algorithms that might support pancreatic cancer diagnosis based on medical imaging modalities.

The oldest study [10] included in this scoping review was published in 2008 and the number of studies has been growing since 40% of the studies were published in 2022 and the first two months of 2023. This shows the current interest in the topic.

In terms of the geographical distribution of the studies, Asia, and North America (have the highest contributions. In this respect, it is possible to infer that the fact that USA and Asian countries such as China and Taiwan lead the research of more effective diagnosis methods for pancreatic cancer, may be due to their having high incidence rates of pancreatic cancer: USA has an incidence rate of 15.9 per 100,000 persons, while Asia presents an incidence rate of 10.2 per 100,000 persons [1].

Table 4. Studies' experimental characteristics.

Ref	Modality	ML Models	Population	Ac	Sens	Spec
[10]	EUS	MLP	46 P + 22 HC	0.861	0.938	0.636
[11]	PET/CT	H-SVM-RF	40 P + 40 HC	0.965	0.952	0.975
[12]	CT	Faster R-CNN	338 P	–	–	–
[13]	CT	DL	100 scans	–	–	–
[14]	CT	CNN	295 P + 256 HC	0.867	0.912	0.858
[15]	CT	CNN	222 P + 190 HC	0.954	0.916	0.982
[16]	CT	DNN	575 P	–	–	–
[17]	CT	XGBoost	728 P + 661 HC	0.888	0.943	0.874
[18]	CT	E2EDL	549 P + 117 HC	0.827	0.868	0.695
[19]	EUS	XGBoost	139 P	–	0.924	0.841
[20]	CT	CNN	400 P + 400 HC	-	0.992	0.975
[21]	CT	SVM	153 P + 159 HC	0.795	0.785	0.806
[22]	CT	DL and NB	112 scans	–	–	–
[23]	CT	NB	36 P + 36 HC	0.893	0.860	0.930
[24]	CT	CNN	157 P	0.950	0.950	0.940
[25]	CT	KNN, SVM, RF, XGBoost	155 P + 265 HC	0.922	0.955	0.903

(*continued*)

Table 4. (*continued*)

Ref	Modality	ML Models	Population	Ac	Sens	Spec
[26]	CT	NB	36 P + 36 HC	0.860	–	–
[27]	CT	RF	85 P + 181 HC	0.935	0.840	0.890
[28]	CT	XGBoost	1217 P + 1593 HC	0.866	0.918	0.822
[29]	CT	CNN	1217 P + 1593 HC	–	0.897	0.928

Notes: Ac. = Accuracy; Sens. = Sensitivity; Spec. = Specificity; P = Patients; HC = Healthy Controls

As seen in Table 4, CT was the preferred medical imaging modality, and it is present in 17 studies. This is a consequence of the fact that CT is the most common pancreatic cancer diagnostic tool [2]. The accurate segmentation of the pancreatic subregions (head, body, and tail) generates images that allow the detection of changes in the pancreas. Quantifying such changes aids in understanding the spatial heterogeneity of the pancreas and assists in the diagnosis and treatment planning of pancreatic cancer [22]. CT provides detailed cross-sectional images, allowing for a comprehensive view of internal structures [22].

Another study focused on the use of PET/CT as the medical imaging modality [11]. Literature suggests that PET/CT doesn't outperform other medical imaging modalities [2]. In turn, it is more expensive, and less available, and involves exposure to radiation [3].

As for EUS, previous studies state that this medical imaging modality has a higher predictive capacity of pancreatic cancer than CT [2, 19]. From the articles reviewed, only two studies used this medical imaging modality with the objective of EUS correctly recognizing a pancreatic mass [10, 19].

In this review, we were expecting to find studies using MRI as the medical imaging modality, as it performs equal to CT, generating detailed images of organs and tissues through radio waves, with the capacity of detecting tumors at an early stage [2]. However, this modality is more expensive, and has some limitations, for example the presence of metal implants in patients [2].

The goal of using ML models to support the diagnosis of pancreatic cancer is to classify medical images to differentiate between benign and malignant lesions or to detect specific diseases from medical imaging studies. This scoping review identified 12 different ML models. The ML model most present is CNN, a data-driven methodology, used for pattern recognition, being able to learn intricate patterns within the dataset, that could be left unnoticed [9]. CNN takes an image as input and transforms it using convolutional layers, pooling layers, and fully connected layers before returning a class-based likelihood of that image [9]. Additionally, it is possible to identify ensemble models (e.g., DL and NB [22]) aiming to improve the performance of the implementations [9].

The numbers of participants of most datasets may raise some concern as they were relatively small. This is particularly important since ML models require a large amount of data to be properly implemented and interpreted. Some studies also use unbalanced datasets of patients and healthy controls, which may have a negative impact on the model's validation.

It is relevant to note that four studies didn't present the performance measures for the ML models, which puts into question the veracity and scalability of the study itself. Analyzing the articles reporting performance values (i.e., accuracy, sensitivity, and specificity) it is possible to conclude that the accuracy varied from 0.795 to 0.965, sensitivity varied from 0.785 to 0.992, and specificity varied from 0.636 to 0.982. However, these high accuracy, sensitivity and specificity values should be carefully interpreted. The researchers worked with datasets in which patients and healthy controls were known from the start, which means that it was possible to optimize the algorithms to obtain the best possible accuracy values.

Furthermore, other performance measures should also be taken into consideration, like the use of confusion matrixes and F1 score, as they provide a more comprehensive assessment of the classifier's performance, instead of the using only the standard accuracy, specificity and sensitivity [6]. Differential studies are required to improve the experience of the machine learning process, such as, randomized control trials, testing the ML models in patients, not knowing the outcome a priori, in order to improve the ability of identifying pancreatic cancer earlier.

In this scoping review we came across some limitations namely the selected databases, the search keywords and the exclusion of all articles not published in scientific journals and not written in English.

5 Conclusion

This review presents the current state of diagnostic tools for pancreatic cancer using ML models. This field of research has had an increasing scientific attention in the last years, indicating that there are still opportunities for future developments.

The results of this review indicated that ML models might be used to support the diagnosis of pancreatic cancer based on medical imaging modalities. However, only retrospective patients' data were analyzed, and patients and healthy controls were known from the start. Therefore, in terms of future developments, there is the need of randomized clinical trials to provide robust evidence to translate ML implementations to clinical practice.

Acknowlegments. This study was carried out within the scope of the course unit Clinical Information Management of the Master's in Clinical Bioinformatics at the University of Aveiro.

References

1. Klein, A.P.: Pancreatic cancer epidemiology: understanding the role of lifestyle and inherited risk factors. Nat. Rev. Gastroenterology Hepatol. **18**(7), 493–502 (2021)
2. Zhang, L., Sanagapalli, S., Stoita, A.: Challenges in diagnosis of pancreatic cancer. World J. Gastroenterol. **24**(19), 2047 (2018)

3. Zhao, Z., Liu, W.: Pancreatic cancer: a review of risk factors, diagnosis, and treatment. Technology in Cancer Research Treatment **19**, 1533033820962117 (2020)
4. National Cancer Institute (n.d.). Cancer Stat Facts: Pancreatic Cancer. https://seer.cancer.gov/statfacts/html/pancreas.html
5. Egawa, S., Toma, H., Ohigashi, H., Okusaka, T., Nakao, A., Hatori, et al.: Japan pancreatic cancer registry; 30th year anniversary: Japan pancreas society. Pancreas, 41(7), 985–992 (2012)
6. Bishop, C.M., Nasrabadi, N.M.: Pattern recognition and machine learning. Springer, New York (2006)
7. Chan, H.P., Samala, R.K., Hadjiiski, L.M., Zhou, C.: Deep learning in medical image analysis. Deep Learning in Medical Image Analysis: Challenges and Applications, pp. 3–21 (2020)
8. Powers, D.M.: Evaluation: from precision, recall and F-measure to ROC, informedness, markedness and correlation. arXiv preprint arXiv:2010.16061 (2020)
9. Murty, M.N., Devi, V.S.: Introduction to pattern recognition and machine learning, vol. 5. World Scientific (2015)
10. Săftoiu, A., et alk.: Neural network analysis of dynamic sequences of EUS elastography used for the differential diagnosis of chronic pancreatitis and pancreatic cancer. Gastrointestinal Endoscopy **68**(6), 1086–1094 (2008)
11. Li, S., Jiang, H., Wang, Z., Zhang, G., Yao, Y.D.: An effective computer aided diagnosis model for pancreas cancer on PET/CT images. Comput. Methods Programs Biomed.. Methods Programs Biomed. **165**, 205–214 (2018)
12. Liu, S.L., et al.: Establishment and application of an artificial intelligence diagnosis system for pancreatic cancer with a faster region-based convolutional neural network. Chin. Med. J. **132**(23), 2795–2803 (2019)
13. Boers, T.G.W., et al.: Interactive 3D U-net for the segmentation of the pancreas in computed tomography scans. Phys. Med. Biol. **65**(6), 065002 (2020)
14. Liu, K.L., et al.: Deep learning to distinguish pancreatic cancer tissue from non-cancerous pancreatic tissue: a retrospective study with cross-racial external validation. Lancet Digital Health **2**(6), e303-e313 (2020)
15. Ma, H., et al.: Construction of a convolutional neural network classifier developed by computed tomography images for pancreatic cancer diagnosis. World J. Gastroenterol. **26**(34), 5156 (2020)
16. Park, S., Cet al.: Annotated normal CT data of the abdomen for deep learning: Challenges and strategies for implementation. Diagnostic interventional Imaging **101**(1), 35–44 (2020)
17. Chen, P.T., et al.: Radiomic features at CT can distinguish pancreatic cancer from noncancerous pancreas. Radiology: Imaging Cancer **3**(4), e210010 (2021)
18. Si, K., et al.: Fully end-to-end deep-learning-based diagnosis of pancreatic tumors. Theranostics **11**(4), 1982 (2021)
19. Tonozuka, R., et al.: Deep learning analysis for the detection of pancreatic cancer on endosonographic images: a pilot study. Journal of Hepato-Biliary-Pancreatic Sci. **28**(1), 95–104 (2021)
20. Wang, Y., Tang, P., Zhou, Y., Shen, W., Fishman, E.K., Yuille, A.L.: Learning inductive attention guidance for partially supervised pancreatic ductal adenocarcinoma prediction. IEEE Trans. Med. Imaging **40**(10), 2723–2735 (2021)
21. Gai, T., Thai, T., Jones, M., Jo, J., Zheng, B.: Applying a radiomics-based CAD scheme to classify between malignant and benign pancreatic tumors using CT images. J. Xray Sci. Technol. **30**(2), 377–388 (2022)
22. Javed, S., Q., et al.: Segmentation of pancreatic subregions in computed tomography images. J. Imaging, **8**(7), 195 (2022)
23. Javed, S., Qet al.: Risk prediction of pancreatic cancer using AI analysis of pancreatic subregions in computed tomography images. Front. Oncol. **12**, 1007990 (2022)

24. Mahmoudi, T., et al.: Segmentation of pancreatic ductal adenocarcinoma (PDAC) and surrounding vessels in CT images using deep convolutional neural networks and texture descriptors. Sci. Rep. **12**(1), 3092 (2022)
25. Mukherjee, S., et al.: Radiomics-based machine-learning models can detect pancreatic cancer on prediagnostic computed tomography scans at a substantial lead time before clinical diagnosis. Gastroenterology **163**(5), 1435–1446 (2022)
26. Qureshi, T.A., et al.: Predicting pancreatic ductal adenocarcinoma using artificial intelligence analysis of pre-diagnostic computed tomography images. Cancer Biomarkers **33**(2), 211–217 (2022)
27. Wang, S., et al.: Compute tomography radiomics analysis on whole pancreas between healthy individual and pancreatic ductal adenocarcinoma patients: Uncertainty analysis and predictive modeling. Technol. Cancer Res. Treatment **21**, 15330338221126869 (2022)
28. Chang, D., et al.: Detection of pancreatic cancer with two-and three-dimensional radiomic analysis in a nationwide population-based real-world dataset. BMC Cancer **23**(1), 1–13 (2023)
29. Chen, P.T., et al.: Pancreatic cancer detection on CT scans with deep learning: a nationwide population-based study. Radiology **306**(1), 172–182 (2023)

Virtual Reality in the Pain Management of Pediatric Burn Patients, A Scoping Review

Joana Santos[1], Jorge Marques[1], João Pacheco[1], and Nelson Pacheco Rocha[2]([envelope]) [ORCID]

[1] Department of Medical Sciences, University of Aveiro, Aveiro, Portugal
{joana27,jorge.m,joao.pacheco}@ua.pt
[2] IEETA, Department of Medical Sciences, University of Aveiro, Aveiro, Portugal
npr@ua.pt

Abstract. Burn care procedures are a major cause of pain, stress and anxiety, and pharmacological methods such as opioids are often insufficient for pain management. This scoping review aimed to understand how virtual reality (VR) might help in the management of pain of pediatric patients suffering from burn wounds. An electronic search was conducted, and thirteen studies were included after the selection process. The studies used various immersive games such as SnowWorld, Dreamland, Virtual River Cruise, Bubbles, ChickenLittle or Need for Speed to reduce pain intensity during the wound dressing change procedures, skin stretching exercises, and hydrotherapy or physical therapy sessions. The results showed the feasibility of using VR to support pain management of burn pediatric patients and pointed for the reduction of pain intensity when using VR, as compared with control groups. Future research should include the customization of the VR solutions to improve the motivational relevance and increase treatment variants, the definition of standardized protocols for clinical implementation, or to determine if it is possible to reduce opioids as a primary outcome.

Keywords: Pediatric Burn Patients · Pain Management · Virtual Reality; Clinical Research · Scoping Review

1 Introduction

Burn wounds are a major health problem globally. The World Health Organization (WHO) estimates that 180 000 deaths every year are caused by burns, the vast majority of those in low- and middle-income countries. Moreover, reports from the American Burn Association show that pediatric burns account for approximately 30% of total burns, with children under the age of five accounting for 17% of all cases [1]. Non-fatal burns constitute the fifth most common cause of non-fatal childhood injuries and are one of the leading causes of morbidity, including prolonged hospitalization, disfigurement, disability, and disability-adjusted life-years lost [2].

Patients with burn injuries suffer physical and psychological distress resulting from the initial trauma of the injuries, and experience both changes related to body image and self-esteem involvement and impacts on lifestyle and towards treatment, including unpleasant procedures that cause pain, suffering, anxiety, and depression. After a

Á. Rocha et al. (Eds.): WorldCIST 2024, LNNS 986, pp. 139–149, 2024.
https://doi.org/10.1007/978-3-031-60218-4_14

traumatic injury, numerous inflammatory cytokines and chemokines are released by the body at the site of injury, causing even adjacent tissues to become painful and sensitive to nociceptive and non-nociceptive stimuli during the healing process [3]. Apart from the background pain, the pain experienced from the procedure of caring for the wounds can be challenging to manage. Many children, especially those with large severe burns, report high pain intensity during burn wound dressing and burn rehabilitation. The burn pain and wound care procedures often elevate the patients' anxiety, which exacerbates their perception of pain. Even though the procedural pain related to burn wound care in children is well acknowledged, it remains undertreated. Children commonly described wound care procedures as the most traumatizing and frightening part of their experience of having a burn and point out that the trauma associated with the process was not only due to the pain but also because they had to witness invasive medical procedures [4].

Even though pediatric wound care procedures are a major cause of pain, stress and anxiety amongst children, physical therapy is a critical component of burn rehabilitation therapy. Early and aggressive physical therapy can help counter decreased range of motion, muscle spasms, limited joint mobility and severe contractures that can develop secondary to burn injury or associated skin grafting [5].

Traditionally, pre-procedure systemic opioid administration has been the treatment method of choice for management of rehabilitation-related pain. Even though opioid analgesics are widely regarded as effective and essential tools for acute pain management, they are associated with major adverse effects, which restrict the dose levels that can be given to the patients and limit the amount of pain reduction from medications alone [6]. The delicate balance between pain relief and opioid requirements is often at the limit of respiratory depression and the occurrence of side effects such as nausea and vomiting. Moreover, tolerance to the medication can also reduce analgesic effectiveness, and patients are at risk of developing opioid addictions. Another problem related to use of these narcotics is their availability, since many non-western countries have little to no availability of opioid analgesics, and even in the United States there are currently shortages of pharmaceutical medical opioid analgesics needed for acute pain control during medical procedures, following the tightening of regulations regarding their use [7, 8]. Hence, there is an increasing interest in non-pharmacological methods as adjuncts to analgesics for optimal pain management [9].

Given that anxiety and other psychological factors can worsen the intensity of the pain felt by the patient, it is reasonable to think that targeting these factors through psychological interventions can reduce pain intensity. In recent years, research studies have recommended using multimodal approaches that do not only combine different pharmacological modalities but also add non-pharmacological components to optimize the relief of procedural pain [10]. Non-pharmacological analgesic techniques include hypnosis, music, toys, video games, parental participation, movies, among others [11–13]. These distraction techniques are based on the gate control theory [14] which hypothesizes that the attention devoted to pain stimuli is limited and pain can be modulated by diverting the attention away from the noxious stimulus. Furthermore, it is hypothesized that distraction techniques that elicit the activation of multiple senses may command the child's attention more than the techniques that only engage one sense, and therefore immersive Virtual Reality (VR) technology poses itself as an exciting solution for non-pharmacological interventions.

VR does not have a clear consensual definition. Several studies have tried to dissect the origins and uses of this terminology [15, 16], but have yet not reached a unifying overall definition due to the various aspects that it encompasses and its multitude of uses, beyond being a distraction method used in health care. For the scope of this work, we consider VR as being a technology that enables its user to view, listen and immerse themselves in an alternate world, through a setup of tri-dimensional (3D) googles, VR helmets or other VR technology that communicate with a computer to generate an artificial, computer-simulated environment, that is then outputted to the users.

In this context, a scoping review was carried out to understand how VR might help in the pain management of pediatric patients suffering from burn wounds, namely by identifying the VR solutions being used and analyzing the clinical outcomes being assessed.

2   Methods

This review was conducted following the Preferred Reporting Items for Systematic Reviews and Meta-Analyses (PRISMA) extension for scoping reviews guidelines [17]. A review protocol was prepared to define (i) search strategy, (ii) inclusion and exclusion criteria, (iii) data extraction, and (iv) synthesis and reporting.

To find relevant studies that have been published in the scientific topic of this review, we selected three primary academic databases: PubMed, Web of Science and Scopus. Boolean queries were prepared to include all the articles that have in their titles, abstract or keywords the expression 'virtual reality' together with the terms 'pain' and 'pediatric'. No date limit was applied to the search.

In terms of studies' eligibility, the inclusion criteria were peer reviewed articles written in the English language reporting studies focused on pediatric patients under the age of 21 years undergoing burn wound care procedures and using VR applications to support pain management.

In turn, the exclusion criteria were references related to pain management of pediatric burn patients without evidence of the use of VR applications, references not written in English, references without abstracts or authors' identification, book chapters or references published in conference proceedings, references reporting on secondary studies such as reviews or surveys, and references whose full texts were not available.

To gather all desired data, a data-charting form was jointly developed by the authors to determine which data to extract. The authors independently charted the data, discussed the results, and continuously updated the data-charting form in an iterative process. Data points pertaining to the following variables were systematically abstracted from each study: publication year, authors, region where the study took place, study design, characteristics of the participants, including number, age and mean Total Body-Surface Area (TBSA), treatment conditions, VR technology used, type of VR intervention, moment of VR use, clinical outcomes, and measurement instruments.

Concerning demographic characteristics, a synthesis was prepared considering the distribution of the studies by geographical areas, according to the regions where they were conducted.

Moreover, in terms of experimental setups, narrative and tabular syntheses were prepared to systematize the studies' design and participants' characteristics, the VR interventions, and the clinical outcomes.

3   Results

3.1   Selection of the Studies

The search on the selected databases was conducted in March 2023 and identified 717 references: (i) 230 from PubMed; (ii) 259 from Web of Science; and (iii) 228 from Scopus.

The results of the selection process of the included studies are presented in Table 1.

Table 1. Selection process.

Excluded	
	334 duplicates
	20 without authors or abstracts
	6 not published in English
	78 secondary studies (reviews, surveys, or guidelines)
	241 excluded by the title and abstract analysis
	25 excluded by the full text analysis
Included	
	13 studies

First, 438 references were removed because they were duplicates, did not have authors' names or abstracts, were not published in English, or reported secondary studies such as reviews, surveys, or guidelines. During the analysis of titles and abstracts 241 references were excluded because they were not related to the topic of the review. Finally, during the full-text analysis, twenty-five references were excluded, namely because they were not focused on pediatric burn patients (e.g., use of VR for venipuncture procedures) or they did not report the use of VR technology. Therefore, thirteen studies were included in this scoping review [18–30].

3.2   Geographical Distribution

Considering the geographical distribution of the studies, United States contributed with six studies [19, 21, 24, 25, 28, 30], Canada with three studies [23, 26, 27], and Australia [20], Brazil [22], Taiwan [18] and Egypt [29] contributed with one study each. The

Canadians studies [23, 26, 27] were conducted by elements of the same research group, while the north American studies were conducted by elements of two different research groups, one from the University of Washington [19, 21, 24, 25], and the other from the Nationwide Children's Hospital [28, 30].

3.3 Studies' Design and Participants

Table 2 presents the characteristics of the studies' design and participants.

Table 2. Studies' design and participants' characteristics.

Reference	Feasibility	Effectiveness	Number of participants	Age (years)	Mean TBSA (%)
[18]	Crossover		8	4–9	-
[19]		RCT	54	6–19	15%
[20]		RCT	41	11–17	5%
[21]	Case study		1	11	36%
[22]	Case study		2	8–9	6%
[23]	Crossover		15	0.2–10	5%
[24]	Crossover		48	7–17	40%
[25]		RCT	50	6–17	44%
[26]	Crossover		38	0.5–7	6%
[27]	Crossover		20	7–17	-
[28]		RCT	90	6–17	3%
[29]		RCT	22	9–16	21%
[30]		RCT	35	5–17	1%

Seven of the studies [18, 21–24, 26, 27] were feasibility studies, either case studies [21, 22] or crossover studies that used the experimental groups as their own controls [18, 23, 24, 26, 27], while the remainder six studies [19, 20, 25, 28–30] were Randomized Control Trials (RCT) aiming to assess the effectiveness of the VR interventions.

All RCT compared the VR intervention group with a control group (standard treatment). Some studies referred standard treatment as the standard practice of the hospital/burn center the study took place in, while others considered the standard treatment an analgesic treatment with injection of morphine and oral administration of tylenol and other drugs.

The number of participants varied from one [21] to ninety [28] and their characteristics were heterogeneous, both in terms of age and the extent of the burn injury (i.e., TBSA, which varied from 1% [30] to 44% [25]). The participants of each study were also quite heterogenous. Some of them had scalds while others were burned (e.g., flame, chemical, electrical or friction burns) with severity levels that varied from first-degree to fourth-degree burns.

3.4   VR Interventions

Table 3 presents the characteristics of the VR interventions and the moments and the moments in which they occurred.

All studies used VR solutions in various forms, including solutions specifically implemented. Moreover, several 3D immersive video games were reported, including SnowWorld [19, 21, 24, 25], Virtual River Cruise [28, 30], Bubbles [23, 26], Dreamland [27], ChickenLittle [20] and Need for Speed [20].

Table 3. VR Interventions.

Reference	Type of VR intervention	Moment of the intervention
[18]	3D immersive VR of an ice-cream factory and auditory senses	During wound dressing change procedure
[19]	VR immersive game (SnowWorld)	During physical therapy sessions
[20]	VR immersive game (Chicken Little and Need for Speed)	During wound dressing change procedure
[21]	VR immersive game (SnowWorld)	During skin stretching exercises
[22]	VR immersive game that simulates a roller coaster	During wound dressing change procedure
[23]	Projector-Based Hybrid VR game (Bubbles)	During hydrotherapy sessions
[24]	VR immersive game (SnowWorld)	During wound dressing change procedure
[25]	VR immersive game (SnowWorld)	During wound dressing change procedure
[26]	Projector-Based Hybrid VR game (Bubbles)	During hydrotherapy sessions
[27]	VR immersive game (Dreamland)	During wound dressing change procedure
[28]	VR immersive game (Virtual River Cruise)	During wound dressing change procedure
[29]	3D immersive VR video	During skin stretching exercises
[30]	VR immersive game (Virtual River Cruise)	During wound dressing change procedure

In terms of the moment of the intervention, most of the studies (i.e., eight studies [18, 20, 22, 24, 25, 27, 28, 30]) the VR interventions occurred during the wound dressing change procedure. Moreover, the VR interventions also occurred during hydrotherapy sessions (i.e., two studies [23, 26]), skin stretching exercises (i.e., two studies [21, 29]) and physical therapy sessions (i.e., one study [19]).

3.5 Clinical Outcomes

The included studies aimed to assess the impact of VR interventions in pain management. Therefore, all the studies included the measurement of pain intensity. Additional clinical outcomes include sensory, affective, and cognitive components of pain [19], anxiety, comfort, and sedation [23], comfort during hydrotherapy sessions [26], pain-related fear, recall of pain and pain-related fear [27], and range of motion [29].

Concerning measurement instruments (Table 4), pain intensity was measured by different scales: Faces Pain Scale [18, 22], Graphic Rating Scale [19, 21, 24, 25, 27]; Visual Analogue Scale [20, 28, 29], Faces, Legs, Activity, Cry, Consolability scale [20, 23, 28] and Numerical Rating Scale [26, 27, 30]. Additional instruments include Behavioral Observational Scale of Comfort Level for Child Burn Victims [23, 27], Ramsay Sedation Scale [23], Pain Catastrophizing Scale for Children [24], Children's Fear Scale [27], and an electronic digital goniometer [29].

Table 4. Measurement instruments.

Reference	Measures
[18]	Faces Pain Scale
[19]	Graphic Rating Scale
[20]	Faces, Legs, Activity, Cry, Consolability scale, and Visual Analogue Scale
[21]	Graphic Rating Scale
[22]	Faces Pain Scale
[23]	Faces, Legs, Activity, Cry, Consolability scale, Modified Smith Scale, Behavioral Observational Scale of Comfort Level for Child Burn Victims, and Ramsay Sedation Scale
[24]	Graphic Rating Scales, and Pain Catastrophizing Scale for Children
[25]	Graphic Rating Scale
[26]	Face, Legs, Activity, Cry Consolability scale, and Numerical Rating Scale
[27]	Graphic Rating Scale, Numerical Rating Scale, Behavioral Observational Scale of Comfort Level for Child Burn Victims and Children's Fear Scale
[28]	Faces, Legs, Activity, Cry, Consolability scale, and Visual Analogue Scale
[29]	Visual analogue scale, and Electronic digital goniometer
[30]	Numerical Rating Scale

Across the RCT studies [19, 20, 25, 28–30], distraction using a VR setup achieved additional analgesia levels when compared with the control treatment, as revealed by the scale scores on self-reported pain and nursing staff observations.

Relative to secondary outcomes, Schmitt and colleagues [19] reported a 3-fold increase in the "fun" experienced by the participants, when allocated to the VR condition group. Similar observations were also reported by Hoffman and colleagues [24, 25].

Despite the positive results, the included studies also concluded about the need of further research to study the customization of the VR solutions to improve the motivational relevance [20] and to increase the treatment variance [19], to establish standardized protocols for clinical implementation, or to determine if is possible to reduce opioids as a primary outcome [28].

4 Discussion and Conclusion

This scoping review identified thirteen studies related to applications of VR to support pain management of pediatric patients suffering from burn wounds. Although the first study was published in 2007, a significant percentage of the studies were published after 2020, which shows the current interest in the topic.

Concerning the geographical distribution, almost 70% of the included studies were conducted by two research groups from United States and a research group from Canada.

The included studies demonstrated the feasibility of using VR to support pain management of burn pediatric patients. Moreover, they also pointed for the reduction of pain intensity when using VR, as compared with control groups. Currently, VR is still widely used as an adjuvant to analgesics, constituting an option for distraction amongst the existing ones.

Concerning VR solutions, various immersive games were used, such as SnowWorld, Virtual River Cruise, Bubbles, Dreamland, ChickenLittle and Need for Speed. Snow-World is becoming progressively obsolete as other games and setups developed to optimize and amplify immersion were developed. Particularly more recent VR games used in health care, such as Dreamland and Virtual River Cruise, which were respectively used by one [27] and two [28, 30] of the included studies, improve the patient's immersion experience through their enhanced interactivity and, therefore, might have a greater impact in the distraction by their users, which is beneficial in the treatment of burn wounds.

VR can be diversified depending on different circumstances, such as the inability to use certain VR equipment such as helmets and rift glasses due to the location, extent, and severity of the burns. This adaptability also improves the comfort experienced by patients, which is of great importance in the type of treatment they undergo. Other situations include the adaptation of VR setups for different age groups, especially regarding the nature and difficulty of the games, which can be a problem and a source of discontent for patients, losing a very important factor provided by this technology, the fun experienced by the patients using it. In addition to the considerable reduction in pain, VR also increases the factor of fun, which is of great relevance in this particular age group to increase patient cooperation. This aspect is also referenced by the clinicians as it diminishes the anxiety felt by the children.

Currently, with the proven use and effectiveness of VR solutions and its variability and adaptability for use in different contexts and circumstances, it is necessary to properly define and structure protocols for the use of diverse types and setups of VR in the treatment of burns in pediatric ages. These protocols should include well-defined methods to facilitate the implementation in clinical practice and the acceptance by the clinicians responsible for providing this care. Moreover, additional RCT are needed to study the possibility of reducing opioids as a primary outcome.

The results of this review must be analyzed considering its limitations. Publication bias may have been generated due to the selected databases, the search queries or the studies' selection process. Particularly, the search query could be enriched with alternative terms to pain such as anxiety, discomfort, and stress.

In conclusion, VR technologies might be effective as an adjunct intervention for pain management in pediatric burn wound health care, offering comfort, and fun for patients while reducing the pain intensity they experience. Additional research is required to explore the customization of the VR solutions to improve the motivational relevance or to increase the treatment variance (e.g., age-specific game content), to establish standardized protocols for clinical implementation and to consider the primary clinical outcome the reduction of opioids.

Acknowledgments. This study was carried out within the scope of the course unit Clinical Information Management of the Master's in Clinical Bioinformatics at the University of Aveiro.

References

1. American Burn Association. 2016 National Burn Repository. American Burn Association (2016)
2. WHO. Burns (2018). https://www.who.int/news-room/fact-sheets/detail/burns
3. Cook, A.D., Christensen, A.D., Tewari, D., McMahon, S.B., Hamilton, J.A.: Immune cytokines and their receptors in inflammatory pain. Trends Immunol. **39**(3), 240–255 (2018)
4. McGarry, S., Elliott, C., McDonald, A., Valentine, J., Wood, F., Girdler, S.: Paediatric burns: from the voice of the child. Burns **40**(4), 606–615 (2014)
5. Procter., F.: Rehabilitation of the burn patient. Indian J. Plastic Surgery: Official Public. Assoc. Plastic Surgeons India **43**(Suppl), S101–S113 (2010)
6. Kessler, E.R., Shah, M., Gruschkus, S., Raju, A.: Cost and quality implications of opioid-based postsurgical pain control using administrative claims data from a large health system: opioid-related adverse events and their impact on clinical and economic outcomes. Pharmacoth. J. Human Pharmacol. Drug Therapy **33**(4), 383–391 (2013)
7. Berterame, S., Erthal, J., Thomas, J., Fellner, S., Vosse, B., et al.: Use of and barriers to access to opioid analgesics: a worldwide, regional, and national study. The Lancet **387**(10028), 1644–1656 (2016)
8. Davis, M.P., McPherson, M.L., Mehta, Z., Behm, B., Fernandez, C.: What Parenteral opioids to use in face of shortages of morphine, hydromorphone, and fentanyl. Am. J. Hospice and Palliative Medicine® **35**(8), 1118–1122 (2018)
9. Retrouvey, H., Shahrokhi, S.: Pain and the thermally injured patient — a review of current therapies. J. Burn Care Res. **36**(2), 315–323 (2015)
10. Birnie, K.A., Noel, M., Chambers, C.T., Uman, L.S., Parker, J.A.: Psychological interventions for needle-related procedural pain and distress in children and adolescents. Cochrane Database Syst. Rev. **2020**(10) (2018)

11. Miller, K., Rodger, S., Bucolo, S., Greer, R., Kimble, R.M.: Multi-modal distraction. Using technology to combat pain in young children with burn injuries. Burns **36**(5), 647–658 (2010)
12. Olsen, K., Weinberg, E.: Pain-less practice: techniques to reduce procedural pain and anxiety in pediatric acute care. Clinic. Pediatric Emergency Med. **18**(1), 32–41 (2017)
13. Rohilla, L., Agnihotri, M., Trehan, S.K., Sharma, R.K., Ghai, S.: Effect of music therapy on pain perception, anxiety, and opioid use during dressing change among patients with burns in india: a quasi-experimental. Cross-over Pilot Study. Ostomy/Wound Manag. **64**(10), 40–46 (2018)
14. Melzack, R., Wall, P.D.: Pain mechanisms: a new theory. Science **150**(3699), 971–979 (1965)
15. Kardong-Edgren, S., Farra, S.L., Alinier, G., Young, H.M.: A call to unify definitions of virtual reality. Clin. Simul. Nurs. **31**, 28–34 (2019)
16. Yoh, M.S.: The reality of virtual reality. In: Proceedings Seventh International Conference on Virtual Systems and Multimedia, pp. 666–674. IEEE (2001)
17. Tricco, A.C., Lillie, E., Zarin, W., O'Brien, K.K., Colquhoun, H., et al.: PRISMA extension for scoping reviews (PRISMA-ScR): checklist and explanation. Ann. Intern. Med. **169**(7), 467–473 (2018)
18. Chan, E.A., Chung, J.W., Wong, T.K., Lien, A.S., Yang, J.Y.: Application of a virtual reality prototype for pain relief of pediatric burn in Taiwan. J. Clin. Nurs. **16**(4), 786–793 (2007)
19. Schmitt, Y.S., Hoffman, H.G., Blough, D.K., Patterson, D.R., Jensen, M.P., et al.: A randomized, controlled trial of immersive virtual reality analgesia, during physical therapy for pediatric burns. Burns **37**(1), 61–68 (2011)
20. Kipping, B., Rodger, S., Miller, K., Kimble, R.M.: Virtual reality for acute pain reduction in adolescents undergoing burn wound care: a prospective randomized controlled trial. Burns **38**(5), 650–657 (2012)
21. Hoffman, H.G., et al.: Feasibility of articulated arm mounted oculus rift virtual reality goggles for adjunctive pain control during occupational therapy in pediatric burn patients. Cyberpsychol. Behav. Soc. Netw. **17**(6), 397–401 (2014)
22. Scapin, S.Q., Echevarría-Guanilo, M.E., Fuculo, P.R.B., Martins, J.C., Barbosa, M.D.V., Pereima, M.J.L.: Use of virtual reality for treating burned children. Rev. Bras. Enferm. **70**, 1291–1295 (2017)
23. Khadra, C., et al.: Projector-based virtual reality dome environment for procedural pain and anxiety in young children with burn injuries: a pilot study. J. Pain Res., 343–353 (2018)
24. Hoffman, H.G., et al.: Immersive Virtual reality as an adjunctive non-opioid analgesic for pre- dominantly Latin American children with large severe burn wounds during burn wound cleaning in the intensive care unit: a pilot study. Front. Human Neurosci. **13** (2019)
25. Hoffman, H.G., Patterson, D.R., Rodriguez, R.A., Peña, R., Beck, W., Meyer, W.J.: Virtual reality analgesia for children with large severe burn wounds during burn wound debridement. Front. Virt. Reality **1** (2020)
26. Khadra, C., Ballard, A., Paquin, D., Cotes-Turpin, C., Hoffman, H.G., et al.: Effects of a projector-based hybrid virtual reality on pain in young children with burn injuries during hydrotherapy sessions: A within-subject randomized crossover trial. Burns **46**(7), 1571–1584 (2020)
27. Le May, S., Hupin, M., Khadra, C., Ballard, A., Paquin, D., et al.: Decreasing pain and fear in medical procedures with a pediatric population (DREAM): a pilot randomized within-subject trial. Pain Manag. Nurs. **22**(2), 191–197 (2021)
28. Xiang, H., Shen, J., Wheeler, K.K., Patterson, J., Lever, K., et al.: Efficacy of smartphone active and passive virtual reality distraction vs standard care on burn pain among pediatric patients: a randomized clinical trial. JAMA Netw. Open **4**(6), e2112082–e2112082 (2021)

29. Ali, R.R., Selim, A.O., Ghafar, M.A.A., Abdelraouf, O.R., Ali, O.I.: Virtual reality as a pain distractor during physical rehabilitation in pediatric burns. Burns **48**(2), 303–308 (2022)
30. Armstrong, M., Lun, J., Groner, J.I., Thakkar, R.K., Fabia, R., et al.: Mobile phone virtual reality game for pediatric home burn dressing pain management: a randomized feasibility clinical trial. Pilot Feasibility Stud. **8**(1), 1–11 (2022)

Defining the "Smart Hospital": A Literature Review

Leonidas Anthopoulos[⊠] [ID], Maria Karakidi, and Dimitrios Tselios[ID]

University of Thessaly, 41500 Larissa, GR, Greece
{lanthopo,makarakidi,dtselios}@uth.gr

Abstract. Smart hospital grassroots date back to the late 20s when the Information and Communication Technologies (ICT) started being adopted by the health sector to digitize its administrative processes. These technologies reduced mistakes, minimized the operational costs, and enhanced the efficiency of hospital management. The primary goal of the ICT utilization is to optimize healthcare delivery using modern data and technologies, improving safety, satisfaction, patient empowerment, clinical outcomes, and performance. Emerging ICT like big data analytics, artificial intelligence, IoT, cloud computing, 5G networks, and blockchain have enabled the integration and interoperability of various data sources and systems, enhancing the efficiency and quality of healthcare services. Moreover, these technologies introduced the terms smart health and smart hospital. The aim of this work in progress is to define smart hospital, explain its context and services, and distinguish it from other emerging approaches (agile, hybrid and green hospital) according to the literature review findings.

Keywords: smart health · smart hospital · digital transformation

1 Introduction

The terms smart health and smart hospital have emerged due to the evolution of the Information and Communication Technology (ICT), which enabled the transformation of the healthcare services from simple digital to enhanced, efficient, combined, and intelligent (Tian et al., 2019). Smart hospital aims to go beyond the optimization of the operational processes that flow within a hospital with the ICT: the quality of the provided healthcare services can be enhanced and become more efficient with the adoption of cutting-edge ICT, like big data analytics, artificial intelligence, IoT, cloud computing, 5G networks, and blockchain.

The aim of this work in progress is to define the "smart" hospital as a critical smart health component, and distinguish it from other adjectives, including "agile", "hybrid" and "green". The definition and clarification of this term are of high importance due to the emergence of smart healthcare, and its prioritization as a pillar of communities' quality of life and of cities sustainability [1]. The authors performed literature review and analyzed their findings with the Preferred Reporting Items for Systematic Reviews and Meta-Analyses (PRISMA) methodology [2] in their attempt to collect the theoretical background and define the term "smart hospital".

© The Author(s), under exclusive license to Springer Nature Switzerland AG 2024
Á. Rocha et al. (Eds.): WorldCIST 2024, LNNS 986, pp. 150–157, 2024.
https://doi.org/10.1007/978-3-031-60218-4_15

The remainder of this article is as follows: Sect. 2 presents the literature review findings and the theoretical background of this article and concludes on the conceptual architecture of smart hospital. Then, Sect. 3 compares smart hospital with other emerging terms, including "agile", "green" and "hybrid". Finally, Sect. 4 contains conclusions and future thoughts for this study.

2   Literature Review

A literature review was performed to collect information about the concept of "smart hospital", its definition and objectives. Search took place in the scientific repositories ScienceDirect and PubMed. The applied keywords were "smart health" "smart hospital", "agile hospital" and "hybrid hospital", which are relative with the purposes of this study. The search process took place in September 2023 and collected articles that were written in English only.

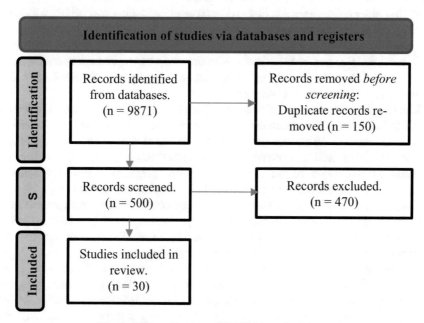

Fig. 1. The application of PRISMA methodology.

After the collection of the extracted pool of articles, a first screening removed duplicates. Then, titles and abstracts were examined and reviewed, and irrelevant articles with the study were excluded from further investigation. Finally, the number of articles was reduced more, after reading all the remaining articles in detail. Since this is a work in progress, the authors followed a "first-in-first-out" methodology, which means that when articles were found repeating similar information to previous ones that were studied, they were ignored. Moreover, the authors used the Mendeley© application to store the studied articles and access selected references that would add value to this study. The Prisma diagram (Fig. 1) depicts the review process, which is still in progress.

2.1 Theoretical Background

No clear and widely accepted definition for smart hospital was found in the articles and the authors attempted to conceptualize the term. The first theoretical evidence of the smart hospital can be traced back to the late '00s century, when ICT the started being adopted in hospitals for administrative purposes, such as electronic medical record registration, billing systems, and scheduling systems, etc. These solutions have helped reducing errors, costs, and inefficiencies in hospital management [3].

According to the collected articles, the primary objective of the smart hospital is to optimize the efficiency and quality of healthcare service delivery in the hospital environment, using data and modern technologies. At the same time, it aims to reduce errors, costs, delays, contamination, and repeated imports. Moreover, its objective is to improve clinical performance and outcomes, while it aims to empower patient, enhance healthcare safety, and improve customer satisfaction. Finally, innovation and research, combined with collaboration and communication efforts are evident.

On the other hand, extensive discussion takes place about smart health, which was raised by the serious challenges that the field of health has faced lately, stemming from the following factors: aging population, escalating healthcare expenditures, an increase in the prevalence of non-communicable diseases and obesity, and substantial research and development costs. The utilization of the emerging digital technologies is crucial for hospitals to address these challenges.

The origins of smart health can be traced back to the early '00s, with the integration of information and communication technology (ICT) in the medical domain for therapeutic purposes. This encompassed the integration of clinical decision support systems, telemedicine, mobile health (mHealth), and electronic health records, among other comparable technologies. In addition to improving outcomes, ICTs have contributed to enhancements in healthcare services' accessibility, quality, and safety.

The digitization of health has been followed by contemporary developments in the ICT, including but not limited to blockchain, big data analytics, AI, and the IoT, which have propelled the emergence of intelligent healthcare facilities and institutions. The advancement of these concepts has been significantly influenced by these technologies. It is currently feasible to integrate and interoperate a vast array of data sources and systems across a variety of healthcare delivery and promotion sectors and levels. Furthermore, they have facilitated the development of novel information and innovations that have substantially enhanced the quality and efficiency of healthcare services, while they promote more precise and individualized medicine.

As a result, the main goal of smart health is to improve the health and well-being of individuals and populations using digital care programs, technologies and services. It also aims to ensure access to health services for all, improve the quality and efficiency of health services, strengthen the prevention of problems and their management, activate personalized care, facilitate work coordination, empower employees and patient education [4].

The above analysis of literature findings was associated with [16, 18, 24] and concluded that *"smart" is the hospital that uses data and technology to improve the efficiency and quality of healthcare service delivery and make medicine more personalized and accurate.*

2.2 Smart Hospital Equipment and Systems

The following components were found to synthesize a smart hospital:

Connected Medical Devices. Mobile devices interconnected with identification systems and clinical information systems, enable remote monitoring of patients, which results to an increasing automation and decision making. Moreover, wearable devices combine advanced sensors, wireless modules, and microprocessors to continuously monitor various physiological indicators of patients, while they reduce energy consumption. These devices are comfort, and they collect data that are combined with health information, which is delivered from other channels.

Identification Systems. An additional feature of smart hospital is the ability to identify persons and objects using biometric scanners. Smart identification systems are used to count and identify both patients and health personnel, authorizing their access to specific areas of the hospital.

Remote Care Equipment. It provides healthcare services outside the hospital, even in isolated areas. Tele-diagnosis and tele-monitoring equipment, equipment for the distribution of medication and tele-health equipment enable these remote services.

Facilities. ICT-based building facilities can support the execution of smart processes. More specifically, energy control systems, temperature sensors, and security management systems which are primarily aimed at patient safety can be installed in a hospital and support its smart transformation.

2.3 Smart Hospital Services

Smart hospitals can improve health monitoring with data collection, which are collected rapidly, they are more reliable and can be transmitted anywhere through fast communication network services. Access to these data is now established even in developing countries [5]. The following ICT-based services can be offered in a smart hospital with the use of data:

IoT-Based Services. IoT technologies connect various objects with sensors and online communication functions, including object recognition, network construction, sensor attachment, and control [6]. They reduce operating costs, enhance treatment effects, minimize diagnostic delays, detect early wear and tear, maximize equipment utilization, improve patient safety, increase building energy efficiency, increase profitability, and improve user experience [7].

Mobile Health Services. They are services provided via mobile devices such as cellphones, tablets, wearables etc. Mobile health record systems, such as personal health records, are designed to enable self-monitoring and management by collecting treatment and examination information from various medical institutions, as well as activity level, weight and blood glucose data collected from smart mobile phones [8]. Previous research related to these records has addressed the usability of mobile systems and their ability to increase patient engagement [9, 10].

AI-Based Services. AI serves as the engine of clinical decision support systems, which assist physicians in making clinical decisions for diagnosis and prescription, thereby improving the effectiveness, efficiency, and safety of treatment. Moreover, AI services based on image recognition are advancing rapidly [11].

Robot Services. They are services that use robots to perform medical and nursing tasks instead of humans. These services cover various areas such as surgery, rehabilitation, and nursing care.

3   A Comparison of Terms

Beyond the emergence of smart hospital, some other approaches to a "future" hospital were also located in literature.

3.1   Agile Hospital

Agile hospitals are embracing the use of digital innovation in their facility management systems and space design. The recent Covid-19 pandemic has dramatically exposed a greater, long-term need in healthcare for this flexibility and adaptability that will enable providers to prepare for and respond to an increasing variety of potential crises and challenges.

Hospitals seeking to develop the flexibility of their buildings can benefit from re-thinking their design and functionality, with particular attention to how spaces can be used effectively in different ways as clinical needs change. They can carefully design the systems and monitoring needed to support this increased flexibility.

Finally, a critical aspect of hospital flexibility is to pay a close attention to the optimal balance between flexibility and cost. The possibility of changing all rooms to negative pressure is possible, but it will add significant costs to the overall project budget, both for the installation of the equipment and the maintenance of the systems. Weighing cost against flexibility and adaptability is necessary. Hospitals that incorporate flexibility into their facilities while managing costs will be better prepared for future changes [12, 15, 19].

3.2   Green Hospital

A green hospital can be defined as a *building that enhances patient well-being, aids the healing process, while using natural resources in an efficient, environmentally friendly manner*. The idea of "green buildings" originated from the persistent efforts of the United Nations to emphasize the importance of "sustainability" in every dimension of human development. The initial effort in this regard was the United Nations Conference on the Human Environment, which took place in 1972 in Stock-holm. The declaration of this conference stated that "Human well-being and economic development throughout the world are affected by one major issue, which is the protection and improvement of the human environment, it is the duty of all governments and the urgent desire of people everywhere world" [13].

Some examples of green hospitals in Europe that have achieved high Building Research Establishment (BREEAM) scores are the hospitals in Brussels, which have a modular design that allows them to adjust their capacity according to demand. Ohio University Wexner Medical Center in Columbus (USA), which has a flexible design that allows it to reconfigure its rooms and equipment for different purposes. The Sheba Medical Center in Tel Aviv (Israel), which has a flexible design that allows it to transform its space and equipment for different purposes, and the Megio Clinic in Rochester (USA), which uses 3D printing to create personalized medical devices, implants, or custom instruments.

The common characteristics of green hospitals are techniques to conserve energy, alternative ways for energy production, waste management, minimizing transportation costs and providing healthy food to both patients and employees [17, 21, 22, 23].

3.3 Hybrid Hospital

The rapid development of technology leads to the digitization of services and processes, which in turn, leads to the reduction or even the elimination of bureaucratic processes. Moreover, the rapid development and evolution of medical technologies leads to less invasive operations and more timely diagnosis, serving prevention more effectively and efficiently.

This transformation of the health sector is based on specific digital technologies that will lead to the provision of high-level healthcare services, from administrative to medical – therapeutic ones. Such digital technologies are synthetic biology and nanotechnology, medical robotics and 3D printing, genomic medicine, blockchain and AI.

An example of a hybrid hospital can be seen in Dordrecht, the Netherlands, the Albert Schweitzer Hospital. The hospital was moved from the center of the old city to its outskirts. Many operational changes were made, the hospital was divided into three levels, hotel facilities and offices, hospital facilities and public space. The modern medical machines used are more efficient, faster, and always connected to achieve a combination of information from various medical fields. A section of the hospital has become mobile. Sensors in clothing and wearable accessories are used by medical staff for rapid diagnosis, medical information is sent to the patient's medical record via wireless connections, while medical staff remain on standby 24 h a day. The idea of a "virtual" i.e., a hybrid Hospital is a goal for the future and creates a new trend in the healthcare field [14].

4 Conclusions and Future Thoughts

This work in progress article addresses the emerging topic of smart hospital. More specifically, it performed a literature review to analyze, define and conceptualize the term. However, the review is still in progress and only some of the collected articles are presented. No clear and widely accepted definition has been found so far in this review - which is still in progress-, but the primary outcomes have generated an understanding of the concept and provided a preliminary definition.

Moreover, smart hospital is based on specific ICT-based technologies and offers several smart medical and care services, both at and beyond the hospital facilities. The smart hospital can differentiate from its competitive terms: the agile hospital emphasizes the use of space; the green hospital addresses the environmental impact; and the hybrid hospital goes beyond intelligence and attempts to regenerate space and improve facilities.

After the completion of the literature review, the authors will define the appropriate transformation strategy of a typical hospital to smart. Theoretical findings will be combined with empirical evidence and interviews with high-level hospital managers.

Acknowledgements. Parts of this article are based on a MSc thesis at the MSc in Agile Management Methods, University of Thessaly, Greece.

References

1. Anthopoulos, L.G.: Understanding Smart Cities: A Tool for Smart Government or an Industrial Trick? Springer, Cham (2017). https://doi.org/10.1007/978-3-319-57015-0
2. PRISMA Homepage. https://www.prisma-statement.org. Accessed 21 Nov 2023
3. Cochrane, C., Hertleer, C., Schwarz-Pfeiffer, A.: Smart textiles in health. In: Smart textiles and their applications, pp. 9–32. Elsevier (2016). https://doi.org/10.1016/B978-0-08-100574-3.00002-3
4. Tian, S., Yang, W., Le Grange, J.M., Wang, P., Huang, W., Ye, Z.: Smart healthcare: making medical care more intelligent. Global Health J. **3**(3), 62–65 (2019)
5. Malik, H., Iqbal, A., Joshi, P., Agrawal, S., Bakhsh, F.I. (eds.): Metaheuristic and evolutionary computation: algorithms and applications. SCI, vol. 916. Springer, Singapore (2021). https://doi.org/10.1007/978-981-15-7571-6
6. Ramson, S.J., Vishnu, S., Shanmugam, M.: Applications of internet of things (IoT)–an overview. In: IEEE 2020 5th International Conference on Devices, Circuits and Systems (ICDCS), pp. 92–95 (2020)
7. Reddy, S., Allan, S., Coghlan, S., Cooper, P.: A governance model for the application of AI in health care. J. Am. Med. Inform. Assoc. **27**(3), 491–497 (2020)
8. Hassan, A., Prasad, D., Khurana, M., Lilhore, U.K., Simaiya, S.: Integration of internet of things (IoT) in health care industry: an overview of benefits, challenges, and applications. In: Taneja, K., Taneja, H., Kumar, K., Selwal, A., Ouh, E.L. (eds.) Data Science and Innovations for Intelligent Systems: Computational Excellence and Society 5.0, 1st edn. CRC Press (2021). https://doi.org/10.1201/9781003132080
9. Bouri, N., Ravi, S.: Going mobile: how mobile personal health records can improve health care during emergencies. JMIR Mhealth Uhealth **2**(1), e3017 (2014)
10. Zhou, L., DeAlmeida, D., Parmanto, B.: Applying a user-centered approach to building a mobile personal health record app: development and usability study. JMIR Mhealth Uhealth **7**(7), e13194 (2019)
11. Park, Y.R., et al.: Managing patient-generated health data through mobile personal health records: analysis of usage data. JMIR mHealth uHealth **6**(4), e9620 (2018)
12. Etemadi, S., Mohammadi, B., Akbarian Bafghi, M.J., Hedayati Poor, M., Gholamhoseini, M.T.: A new costing system in hospital management: time-driven activity based costing: a narrative review. Evid. Based Health Policy Manage. Econ. **2**(2), 133–140 (2018)
13. Sohn, L.B.: Stockholm declaration on the human environment. Harvard Int. Law J. **14**(3), 423–515 (1973)

14. Wagenaar, C. (ed.): The Architecture of Hospitals. NAI Publishers, Rotterdam (2006)
15. Anesi, G. L., Lynch, Y., Evans, L.: A conceptual and adaptable approach to hospital prepared-
    ness for acute surge events due to emerging infectious diseases. Crit. Care Explor. 2(4), e0110
    (2020)
16. Holzinger, A., Röcker, C., Ziefle, M.: From smart health to smart hospitals. Smart Health:
    Open Problems and Future Challenges, 1–20 (2015)
17. Lennox, L., Linwood-Amor, A., Maher, L., Reed, J.: Making change last? Exploring the value
    of sustainability approaches in healthcare: a scoping review. Health Res. Pol. Syst. 18(1), 1–24
    (2020)
18. Nasr, M., Islam, M.M., Shehata, S., Karray, F., Quintana, Y.: Smart healthcare in the age of AI:
    recent advances, challenges, and future prospects. IEEE Access 9, 145248–145270 (2021)
19. Ndayishimiye, C., et al.: Associations between the COVID-19 pandemic and hospital infras-
    tructure adaptation and planning—a scoping review. Int. J. Environ. Res. Public Health 19(13),
    8195 (2022)
20. Papadopoulos, A.M.: Energy efficiency in hospitals: Historical development, trends and per-
    spectives. Energy Performance of Buildings: Energy Efficiency and Built Environment in
    Temperate Climates, 217–233 (2015)
21. Prada, M., et al.: New solutions to reduce greenhouse gas emissions through energy efficiency
    of buildings of special importance–hospitals. Sci. Total. Environ. 718, 137446 (2020)
22. Stevanovic, M., Allacker, K., Vermeulen, S.: Hospital building sustainability: the experience
    in using qualitative tools and steps towards the life cycle approach. Procedia Environ. Sci. 38,
    445–451 (2017)
23. Vergunst, F., Berry, H.L., Rugkåsa, J., Burns, T., Molodynski, A., Maughan, D.L.: Applying
    the triple bottom line of sustainability to healthcare research—a feasibility study. Int. J. Qual.
    Health Care 32(1), 48–53 (2020)
24. Zhao, W., Luo, X., Qiu, T.: Smart healthcare. Appl. Sci. 7(11), 1176 (2017)

Deep Learning for Healthcare:
A Web-Microservices System Ready
for Chest Pathology Detection

Sebastián Quevedo[1,3]([✉]), Hamed Behzadi-Khormouji[2], Federico Domínguez[1], and Enrique Peláez[1]

[1] Electrical and Computer Engineering, Escuela Superior Politécnica del Litoral - ESPOL University, Guayaquil, Ecuador
asqueved@espol.edu.ec
[2] University of Antwerp, imec-IDLab, Antwerp, Belgium
[3] Universidad Católica de Cuenca, XR-LAB, Cuenca, Ecuador

Abstract. The automation of medical diagnosis has accelerated thanks to the integration of artificial intelligence (AI), particularly in interpreting pathologies in chest X-rays. This study presents a web microservices system that uses a deep learning model to classify thoracic pathologies. The system improves clinical decision-making by providing visual aids, including heat maps for model explainability and a comprehensive set of medical image manipulation tools. The back-end, developed using a microservices architecture, ensures robust data management, secure user authentication, and efficient AI model integration. The results highlight the system's accuracy in detecting pathologies with an average AUC of 0.89, an easy-to-use interface, and the transformative impact of AI explainability in clinical settings.

Keywords: Deep Learning · Chest X-rays · AI Explainability · Healthcare Technology · Microservices Architecture

1 Introduction

The landscape of medical diagnostics has been significantly transformed by emerging technologies, with diagnostic imaging standing at the forefront of this revolution [1–3]. Central to this domain, chest X-rays are crucial in detecting and managing a spectrum of lung and cardiovascular pathologies [4]. However, their interpretation demands specialized expertise, and the inherent subjectivity of human analysis can lead to diagnostic discrepancies and errors [5].

Integrating AI into medical diagnostics has considerably enhanced diagnostic accuracy and consistency. This integration is particularly evident in the realm of chest X-ray interpretation, where AI technologies, notably Convolutional Neural Networks (CNNs) such as DenseNet121, have shown considerable capability in providing detailed and consistent interpretations [6–11]. The effectiveness of these AI models is primarily attributed to the extensive public databases available, including Chexpert, NIH, Padchest, and MIMIC [12–14]. These databases have played an important role in the training and validation processes of the AI models, contributing significantly to the development of robust and reliable AI systems for diagnostic purposes. AI technologies have established thorough validation protocols through these resources, enhancing the trustworthiness and applicability of AI-driven diagnostic methods in medical practice.

Despite these academic strides, the transition of AI models from research to real-world clinical settings still needs to be faster and fraught with challenges [15]. Technical barriers, system adaptability, scalability issues, data security concerns, and integration complexities with existing medical infrastructures are major impediments to these technologies' practical deployment.

To address these challenges, the microservices architecture has been emerged as a viable framework, offering modular, scalable, and efficient solutions apt for the data-intensive nature of the medical field [16]. This study aims to delineate the implementation and assessment of a microservices-based web information system designed to bridge the gap between research innovation and clinical application. The proposed system integrates encryption technologies, authentication mechanisms, and AI techniques to facilitate reliable pathology diagnosis in chest X-rays. A feature of this system is the incorporation of AI explainability techniques, enhancing the transparency and trustworthiness of the AI models' decision-making processes. This study presents a tool that seamlessly integrates into clinical workflows, augmenting the diagnostic process with objective and substantiated insights.

The subsequent sections of this article will delve into related work, methodology, results, and conclusions, outlining the comprehensive journey from concept to implementation of this innovative AI-integrated diagnostic tool.

2   Related Work

In the realm of AI applied to chest X-ray analysis, several studies have laid the groundwork for advancements in this field [17–20]. This section explores these studies and discusses how our research builds upon and contributes to these works.

The CheXpedition study [17] explores the generalization challenges of AI algorithms in chest X-ray analysis. It mainly evaluates the models' efficacy in Tuberculosis (TB) detection and pathology identification from varied datasets, emphasizing the need for models to generalize across diverse institutional and disease contexts. This study aims to enhance the diagnostic process by integrating a flexible microservices architecture and implementing advanced AI models like DenseNet-121 to enhance adaptability and generalization in different clinical environments.

The Chester study [18] introduces a web-based system for chest X-ray diagnostic predictions, emphasizing out-of-distribution detection and disease prediction. This research complements that initiative with a similar web-based interface, further enhanced by incorporating AI explainability and robust data security measures. These additions support the system's reliability and confidence in AI-driven diagnostic outcomes.

[19] focuses on developing a mobile health (M-HEALTH) system leveraging deep learning to detect COVID-19 in chest X-rays. A key aspect of this system is its incorporation of robust data security approaches, ensuring the confidentiality and integrity of medical data. Our research not only supports COVID-19 detection but also extends its capabilities to identify other pathologies through AI. By integrating robust data security protocols, our system aligns with the increasing demand for secure data handling in healthcare technologies.

Lastly, the work of Pham et al. [20] focuses on the accuracy and explainability of AI in interpreting chest radiographs. Aligning with these objectives, our research leverages advanced AI explainability techniques to promote a more precise understanding and trust in AI-based diagnostics. Including methods like GradCAM in our system provides a more transparent view of the AI decision-making process, aiming to foster greater acceptance and confidence in AI-assisted diagnostics among medical professionals.

3   Methods

A diagnostic system for analyzing chest X-rays has been developed, integrating various technological and methodological components. The Methods section provides a detailed overview of the system's structure and capabilities, covering key aspects such as imaging technology, data processing algorithms, Deep Learning techniques for image analysis, microservices approach, and user interface design. Each of these elements contributes to the overall functionality and efficiency of the system, as detailed in the respective subsections. Each subsection deals with a specific aspect of the system, as shown in Fig. 1.

3.1   Front-End Development

Framework and Technologies: The front-end interface leverages Angular[1], a robust framework known for its dynamic data binding and modular component-based architecture. This choice ensures a scalable and maintainable user interface. HTML5[2] and CSS[3] are utilized for structuring and styling the web pages, providing a modern and responsive design.

Medical Image UI: A key feature in the front-end is the integration of Cornerstone.js, a specialized library specifically designed for medical imaging. It facilitates the rendering and manipulating of medical images directly in the web browser without requiring specialized software.

---

[1] https://angular.io/.

[2] https://www.w3.org/TR/2014/REC-html5-20141028/.

[3] https://www.w3.org/Style/CSS/Overview.en.html.

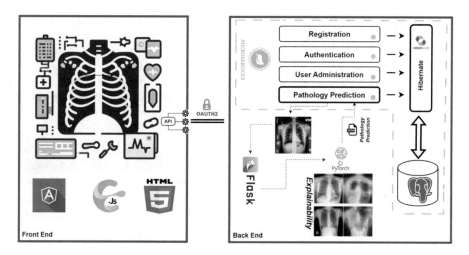

Fig. 1. Architectural AI-Integrated for Chest X-Ray Pathology Detection: front-end in Left Box, back-end in Right Box

Image Manipulation Capabilities: The system allows users to interact with medical images extensively, including following features.

- Zooming in and out of images for detailed examination.
- Adjusting color gradients to enhance image clarity.
- Performing annotations for marking areas of interest or concern.

These functions are critical for medical professionals to interpret chest X-ray images accurately.

Responsive Design: the interface is responsive, ensuring the application is accessible and user-friendly across various devices and screen sizes. This adaptability is crucial for medical professionals accessing the system from different devices.

API Calls: In the AI-integrated diagnostic system's front-end development, The HTTP Client is employed for secure API interactions with the back-end, which is crucial for data retrieval and submission operations. Integral to this process is the implementation of OAuth 2.0, which enhances security by managing authentication and authorization for API access. OAuth 2.0 generates access tokens during user authentication, which is essential for validating and authorizing API calls.

These tokens are embedded in the HTTP headers, ensuring all data transmissions are encrypted, and access is only restricted to authorized users. Additionally, the system includes robust mechanisms for securely storing and managing these tokens within the client's session, maintaining session integrity, and safeguarding sensitive medical data and user information.

3.2   Back-End Development

The back-end architecture of the diagnostic system is designed to support the advanced functionalities required for efficient medical image processing and analysis. This section outlines the key components and functionalities of the back-end development. For a detailed overview of each microservice and its objectives, see Table 1.

Microservices Architecture: The system is structured around a microservices architecture, facilitating scalability, maintainability, and independent updating of various functionalities. Each microservice is focused on a specific task, ensuring a high degree of specialization and efficiency.

These microservices interact through well-defined application programming interfaces (APIs), enabling a seamless data flow and integration of various system components.

Table 1. Overview of Key Microservices

Microservice	Objective
Registration	Simplifies the user registration process by enabling easy entry of personal information
Authentication	Manages user information and roles, using username and password authentication. Leverages Oauth2.0 for token generation, which controls access to microservices based on user roles
Role Management	Enables administrators to effectively oversee and adjust user roles within the system, impacting user permissions and access levels across the platform
User Data Management	Provides users with the capability to update their personal information, such as name, email, phone number, and other pertinent details, ensuring accuracy and an enhanced user experience
Pathology Prediction	Utilizes advanced algorithms and data analytics to predict potential health pathologies in users, based on their entered health data. This service aims to provide early warnings and facilitate preventive healthcare measures

Security Implementation:

- Authentication and Authorization: A robust security framework is integral to the back-end architecture. Utilizing OAuth 2.0, the system ensures secure authentication and authorization processes, safeguarding sensitive medical data and user information.
- Data Encryption: To enhance security, we encrypt sensitive data, ensuring the confidentiality and integrity of patient and diagnostic information.

Data Management The system's back-end uses a framework called Hibernate, an Object-Relational Mapping (ORM) tool. This helps to interact with the database quickly and efficiently. Using this approach makes it possible to represent data object-oriented, which makes data management tasks a lot easier. The system utilizes a Postgres database known for its robustness and compatibility with the ORM framework.

3.3   AI Model Integration and Pathology Prediction

Incorporating the DenseNet-121 CNN into the diagnostic system represents a significant advancement in medical imaging, particularly in the analysis of chest X-rays. This model noted for its performance in Stanford University's CheXpert competition, was further refined using the Deep AUC Maximization approach (DAUC) [21], enhancing its capability for accurate and efficient pathology detection.

Training Process: The model underwent a specialized training regimen using the DAUC approach. This training was designed to optimize the model's performance for detecting pathologies in chest X-rays, enhancing its diagnostic accuracy and reliability.

Alongside DAUC transfer learning techniques were employed to tailor the model to the system-specific dataset, ensuring its robust performance in the specialized task of chest X-ray analysis.

Microservice Architecture for Model Integration: A Flask[4] microservice was developed to integrate the DenseNet-121 model with the diagnostic system's back-end. This microservice manages the data processing and exchange, facilitating efficient communication between the AI model and the rest of the system.

The Flask microservice enables the seamless processing of chest X-ray images, allowing the DenseNet-121 model to perform its diagnostic functions and relay predictions back to the system's back-end.

Diagnostic Data Processing and Reporting: Within the Flask microservice, chest X-ray images are preprocessed and optimized for analysis by the DenseNet-121 model. The model, trained with Deep AUC Maximization, processes these images to detect pathologies. The back-end assembles the model's predictions into comprehensive diagnostic reports, which include detailed information on the identified pathologies, contributing significantly to the clinical decision-making process.

---

[4] https://flask.palletsprojects.com/en/3.0.x/.

Explainability Framework: The system uses the Gradient-weighted Class Activation Mapping (GradCAM) [22] technique to provide visual explanations for the AI model's predictions by highlighting specific areas in chest X-ray images that influenced the model's decision-making. These explainability techniques are seamlessly integrated with the DenseNet-121 Convolutional Neural Network, ensuring that the AI model's output is accurate and interpretable. The Grad-CAM integration allows for the generation of heatmaps overlaid on the original chest X-ray images. These heatmaps visually represent the regions of interest that the model focuses on while making predictions, offering clinicians intuitive insights into the model's analytical process. By providing these visual explanations, the system aids healthcare professionals in understanding the rationale behind the AI model's pathology predictions. This feature is instrumental in building trust in AI-assisted diagnostics and facilitating informed clinical decision-making.

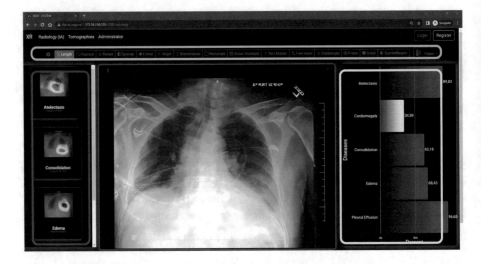

Fig. 2. Web-Microservices System UI

4   Results

The proposed web microservices for chest pathology detection are illustrated in Fig. 2, offering a suite of features for enhanced clinical analysis. In the realm of Model Explainability it includes heatmaps, highlighted within a red rectangle, which overlay chest X-ray images to pinpoint critical areas that influence the AI model's predictions. The Image Manipulation Toolbar, encased in a blue rectangle, offers a comprehensive set of tools for medical image manipulation, enhancing the ability of clinicians to interact effectively with the X-ray images. Additionally, the Predictive Probabilities Display, marked by a yellow rectangle, graphically presents the predictive probabilities for each identified pathology, enabling clinicians to gauge the AI model's diagnostic confidence quantitatively. This integrated approach aims to augment clinical decision-making with intuitive, informative visual aids.

4.1 Performance Evaluation of the Model

The DenseNet-121 CNN, applying the DAUC approach, was evaluated using the CheXpert Dataset [12]. This assessment aimed to ascertain the model's capability in interpreting chest X-rays accurately. In the testing phase, the DenseNet-121 CNN model was evaluated against a dataset derived from CheXpert. The primary metric used for this evaluation was the Area Under the Curve (AUC). The model demonstrated an average AUC of 0.89, indicating its proficiency in the given task. Further insights into the model's performance were obtained by examining the AUC scores for individual pathologies, as depicted in Fig. 3. This figure illustrates the model's diagnostic performance across various pathologies, offering a detailed view of its effectiveness.

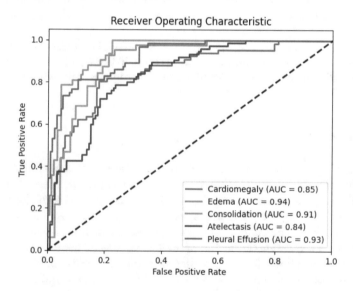

Fig. 3. ROC Curve Analysis: Assessing the Predictive Accuracy of the Classification Model

4.2 AI Explainability Integration

Integrating AI explainability techniques within the diagnostic system represents a critical advancement in making AI-driven medical diagnostics more transparent and clinically relevant. This section elucidates how explainability mechanisms are employed in the AI model's decision-making processes, particularly in the analysis of chest X-rays. This process, as exemplified in Fig. 4.

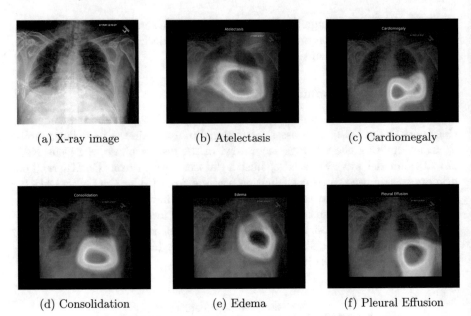

(a) X-ray image   (b) Atelectasis   (c) Cardiomegaly

(d) Consolidation   (e) Edema   (f) Pleural Effusion

Fig. 4. Visualization of Explainability Process: Sections labeled from 'a' to 'f.' Section 'a' displays the X-ray image under analysis. Sections 'b' through 'f' illustrate the Grad-CAM, highlighting the areas of interest in the X-ray image that contribute most significantly to the model's predictions.

5   Discussion

The suite of features offered by the web microservices enhances clinical analysis; integrating the model explainability through heatmaps, provides a helpful tool for clinicians. These heatmaps overlay on chest X-ray images to pinpoint critical areas, thus clarifying the AI model's decision-making process. This feature not only adds a layer of transparency to the use of AI, but also could aid clinicians in understanding the rationale behind specific diagnoses. This interactivity could be critical, as it allows for a more detailed and nuanced examination of the images, potentially leading to more accurate interpretations and diagnoses.

The application of the DAUC approach and its evaluation using the CheXpert Dataset shows the model's capability in interpreting chest X-rays with an average AUC of 0.89. This level of accuracy and the detailed breakdown of the AUC scores for individual pathologies offers an in-depth understanding of the model's performance across a spectrum of conditions, highlighting its comprehensive diagnostic capabilities.

Integrating AI explainability techniques within the diagnostic system not only enhances the transparency of AI models but also elevates their clinical relevance. By elucidating the AI's decision-making processes in the analysis of chest X-rays, the system bridges the gap between advanced AI technologies and practical clinical applications.

6   Conclusions

This study presents the development and implementation of a web-based microservices system for the enhanced analysis of chest X-rays, leveraging the advanced capabilities of artificial intelligence (AI), specifically the DenseNet-121 Convolutional Neural Network. Integrating this technology within a user-friendly and efficient microservices architecture has substantially enhanced the diagnostic process, offering a novel approach to chest pathology detection.

The integration of explainability methods, mainly through heatmaps, has provided an intuitive and transparent understanding of the AI model's decision-making process. This has been key in improving clinician trust and comprehension of AI-driven diagnostics.

The user interface, equipped with comprehensive image manipulation tools and a predictive probabilities display, has considerably enriched the user experience. Clinicians have been able to interact with medical images more effectively.

The back-end system, developed using a microservices architecture, played a crucial role in the system's overall performance. It facilitated seamless data management, secure user authentication, and efficient interaction with the AI model. This robust back-end development has ensured the system's stability, scalability, and security.

7   Future Works

Model Adjustment with Additional Data Sets: To enhance the model's generalization, plans are in place to incorporate broader and more varied data sets, including MIMIC-CXR [14], NIH [12] and PadChes [13]. This strategy is intended to provide a comprehensive array of data for training and validation, thus ensuring the model's applicability in various clinical settings and patient demographic profiles, thereby strengthening the model's generalization.

Multimodal Data Integration: Combining clinical text with chest radiography images offers a more complete perspective of patient data, which could improve diagnostic accuracy. This multimodal approach aligns with a comprehensive treatment in patient diagnosis.

Implementation of Explainability Techniques in AI: To enhance the system's transparency and ongoing development, various AI explainability techniques will be implemented. These techniques aim to increase the system's accuracy and give physicians a more detailed understanding. Integrating these explainability methods into the system's user interface will offer understandable tools for comprehending AI reasoning.

Expansion to Other Medical Imaging Modalities: The system will be expanded to include functionalities for different types of medical imaging, such as MRI and CT scans. This expansion seeks to increase the system's utility for diverse diagnostic imaging requirements.

Acknowledgment. The Ecuadorian government has supported the first author under a SENESCYT scholarship.

References

1. Lodwick, G.S., Keats, T.E., Dorst, J.P.: The coding of roentgen images for computer analysis as applied to lung cancer. Radiology **81**, 185–200 (1963)
2. Zakirov, A., Kuleev, R., Timoshenko, A., Vladimirov, A.: Advanced approaches to computer-aided detection of thoracic diseases on chest X-rays. Appl. Math. Sci. **9**, 4361–4369 (2015)
3. Qin, C., Yao, D., Shi, Y., Song, Z.: Computer-aided detection in chest radiography based on artificial intelligence: a survey. Biomed. Eng. Online **17**, 113 (2018)
4. Rajpurkar, P., et al.: Deep learning for chest radiograph diagnosis: a retrospective comparison of the CheXNeXt algorithm to practicing radiologists. PLoS Med. **15**, e1002686 (2018)
5. Brady, A.P.: Error and discrepancy in radiology: inevitable or avoidable? Insights Imaging **8**, 171–182 (2017)
6. Pati, S., et al.: Author correction: federated learning enables big data for rare cancer boundary detection. Nat. Commun. **14**, 436 (2023)
7. Quevedo, S., Domıngez, F., Pelaez, E.: Detection of pathologies in X-Ray chest images using a deep convolutional neural network with appropriate data augmentation techniques. In: 2022 IEEE ANDESCON, pp. 1–6 (2022)
8. Quevedo, S., Domıngez, F., Pelaez, E.: Detecting multi thoracic diseases in chest X-Ray images using deep learning techniques In: 2023 IEEE 13th International Conference on Pattern Recognition Systems (ICPRS), pp. 1–7 (2023)
9. Jaiswal, A.K., et al.: Identifying pneumonia in chest X-rays: a deep learning approach. Measurement **145**, 511–518 (2019)
10. Ebrahimighahnavieh, M.A., Luo, S., Chiong, R.: Deep learning to detect Alzheimer's disease from neuroimaging: a systematic literature review. Comput. Methods Programs Biomed. **187**, 105242 (2020)
11. Huang, G., Liu, Z., Van Der Maaten, L., Weinberger, K.Q.: Densely connected convolutional networks. In: Proceedings of the IEEE Conference on Computer Vision and Pattern Recognition, pp. 4700–4708 (2017)
12. Wang, X., et al.: ChestX-ray8: hospital-scale chest X-ray database and benchmarks on weakly-supervised classification and localization of common thorax diseases. In: Proceedings of the IEEE Conference on Computer Vision and Pattern Recognition, pp. 2097–2106 (2017)
13. Bustos, A., Pertusa, A., Salinas, J.-M., de la Iglesia-Vayá, M.: PadChest: a large chest X-ray image dataset with multi-label annotated reports. Med. Image Anal. **66**, 101797 (2020)
14. Johnson, A.E., et al.: MIMIC-CXR-JPG, a large publicly available database of labeled chest radiographs (2019). arXiv preprint arXiv:1901.07042
15. Mandreoli, F., Ferrari, D., Guidetti, V., Motta, F., Missier, P.: Real-world data mining meets clinical practice: research challenges and perspective. Front. Big Data **99** (2022)

16. Quevedo, S., Merchán, F., Rivadeneira, R., Dominguez, F.X.: Evaluating apache OpenWhisk-FaaS. In: 2019 IEEE Fourth Ecuador Technical Chapters Meeting (ETCM), pp. 1–5 (2019)
17. Rajpurkar, P., et al.: CheXpedition: Investigating generalization challenges for translation of chest X-ray algorithms to the clinical setting (2020). arXiv preprint arXiv:2002.11379
18. Cohen, J. P., Bertin, P., Frappier, V.: Chester: A web delivered locally computed chest X-ray disease prediction system (2019). *arXiv preprint*
19. Delgado, J., Clavijo, L., Soria, C., Ortega, J., Quevedo, S.: M-HEALTH system for detecting COVID-19 in chest X-Rays using deep learning and data security approaches. In: International Congress on Information and Communication Technology, pp. 73–86 (2023)
20. Pham, H.H., Nguyen, H.Q., Nguyen, H.T., Le, L.T., Khanh, L.: An accurate and explainable deep learning system improves interobserver agreement in the interpretation of chest radiograph. IEEE Access **10**, 104512–104531 (2022)
21. Yuan, Z., Yan, Y., Sonka, M., Yang, T.: Large-scale robust deep AUC maximization: a new surrogate loss and empirical studies on medical image classification. In: Proceedings of the IEEE/CVF International Conference on Computer Vision, pp. 3040–3049 (2021)
22. Selvaraju, R.R., et al.: Grad-CAM: visual explanations from deep networks via gradient-based localization. In: Proceedings of the IEEE International Conference on Computer Vision, pp. 618–626 (2017)

Risk Factors in the Implementation of Information Systems in a Federal University Hospital

Eliane Cunha Marques$^{(\boxtimes)}$ (iD), Simone B. S. Monteiro (iD), Viviane V. F. Grubisic (iD), and Ricardo Matos Chaim (iD)

University of Brasilia, Brasilia, DF 70910-900, Brazil
eliane_marques@hotmail.com, {viviane.grubisic,ricardoc}@unb.br

Abstract. This article presents a comparative study between the risk factors in the implementation of information systems found in the literature and those identified in the implementation of the Hospital Management Application of University Hospitals - AGHU, in a federal university hospital. Information Technology has evolved with regard to complex systems – Information Systems – requiring studies to assess the risks involved in the implementation of hospital systems. The field research involved specialists involved specialists involved in the implementation of the AGHU modules. The results confirmed that risk factors occur in this environment, showing consensus among experts for the following risk factors: difficulties in using the system, lack of resources, users' resistance to changes, and emergence of new requirements. The process of implementing systems, despite the application of risk management techniques, is still prone to failures, and it is necessary to constantly identify, assess its probability of occurrence, estimate its impacts, treat and establish a contingency plan in case the problem actually occurs.

Keywords: Risk · Risk Factors · Hospital Information Systems · implementation of systems · Federal University Hospital

1 Introduction

Information systems are important tools in the strategic management of companies' knowledge. They provide agility, versatility and availability of information, in addition, they must also ensure that only authorized people have access to the database and identify the person responsible for the available information [1].

Health information systems (HIS) collect, process, store, and distribute information to support the decision-making process and assist in the control of health organizations. Thus, they bring together a set of data, information and knowledge used in the sector, to support the planning, improvement and decision-making process of the multiple professionals in the area involved in the care of patients and users of the system [2].

In this context, the present study stands out by addressing the theme of risk identification in the implementation phase of Hospital Information Systems, from the perspective of Federal University Hospitals.

In view of this situation, the following research question arises: are the risk factors in the implementation phase of Information Systems, identified in the literature, compatible with the risk factors in the implementation phase of the Management Application for University Hospitals (AGHU) in a Federal University Hospital?

To answer the question, the objective of this study is to identify the main risk factors in the implementation phase of the Management for University Hospitals (AGHU) application, in a Federal University Hospital, in order to compare them with the risk factors raised by Santos et al. (2020) in a literature review article [4].

2 Related Work

In order to identify the works related to the proposed theme, searches were carried out in the *Web of Science* and *Google Scholar databases*, with the following descriptors used in the *bibliographic search* string: *"systems implementation" AND "*hospital*" OR "information technology"* AND "risk" OR "risk factors*" AND "hospital information system"* AND *"federal university hospital"*, retrieving a total of 78 publications, in the *Web of Science* database, distributed in categories in which the articles are classified, with emphasis on "Health Sciences Services", which has the most publications.

The articles are classified into the following categories, with their respective quantities: Health Sciences Services, with 22 publications; Medical Informatics, with 22 publications; Computer Science Information Systems, with 11 publications; Interdisciplinary Applications of Computer Science, with 7 publications; Health policy services, with 7 publications; Public Environmental Occupational Health, with 7 publications; Methods of Computer Science Theory, with 6 publications; Information Science and Library Science, with 5 publications; Surgery, with 5 publications; Electronic Electrical Engineering, with 4 publications; and Surgery, with 5 publications. This Web *of Science* categorization scheme can be seen in Graph 1, below:

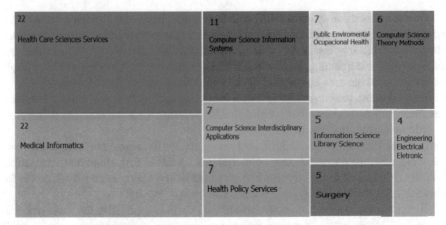

Graph 1. Treemap chart with Web of *Science categories*

The most cited authors were: Ahmad, rablah; Ismail, zuraini; and Samy, narayana ganthan.

3 Theoretical Framework

3.1 Information Technology Risk

Information Technology has evolved with regard to complex systems – Information Systems – in such a way that studies are necessary to evaluate its use in terms of the risks involved in the implementation of hospital systems [3, 5].

Risk is the probability of a certain threat occurring, exploiting vulnerabilities in an asset or a set of assets, aiming to harm the organization, which may cause an incident [6].

Risk can also be understood as an uncertain event or condition that can have a positive or negative effect on the company's objectives. It is observed that many issues can invalidate the entire *software* process. Risk is a potential problem, which may or may not occur. Regardless of the outcome, it is advisable to identify it, assess its likelihood of occurrence, estimate its impact, and establish a contingency plan in case the problem actually occurs [15, 16].

3.2 Hospital Information System

Information Systems consist of a set of computerized technologies that integrate information, people, and organizational processes [8].

Hospital information systems consist of integrated hospital information processing systems that support health activities at the operational, tactical, and strategic levels [9–12], the International Association of Medical Informatics, *International Medical Information Association* (*IMIA*), recommended that patient care should become the main focus of hospital information systems [7].

Patient management systems, considered as the new generation of systems to integrate a hospital information system, are more recent and originated in the development of patient record systems.

Complex and recognized hospital information systems, such as DHCP (*Decentralized Hospital Computer Program*), now installed in dozens of health facilities of the Veterans Administration and the U.S. Department of Defense, had as its basic and initial module the patient registration system [18].

The same was true of the *Division d'Informatique Hôpital Genève* (DIOGENE), a complex information system housed in a 1,600-bed hospital at the University of Geneva and also noteworthy is the famous information system *The Computer Stored Ambulatory Record*, developed by Barnett and his colleagues at the Computer Science Laboratory of Massachusetts General Hospital, and which is based on the patient registration system [17].

3.3 Implementation of the Management Application for University Hospitals - AGHU

In Brazil, the Federal University Hospitals have the Management Application for University Hospitals (AGHU), which is a hospital management system focused on the patient, with the function of assisting in the management and standardization of the care and administrative activities of the Federal University Hospitals, thus allowing the creation of national indicators and, consequently, improvement programs for hospitals [8].

The AGHU is a hospital management system, constituted in a modular format, covering the care processes, administrative processes, operational controls, workflows and analysis of information and indicators of the hospital. Its main objective is to support the standardization of care and administrative practices of Federal University Hospitals and to allow the creation of national indicators, facilitating the creation of common improvement programs for all these hospitals [8]. The updates of the AGHU have been occurring continuously and with this its functionalities are extended and new modules are incorporated.

The adoption of the modules by the Federal University Hospitals is processed individually and depends on the activities developed in the hospital [8].

Currently, AGHU is in its version 10 and consists of 19 modules, as can be seen in Fig. 1, below:

The implementation of systems consists of carrying out the necessary activities for the system to be installed for its use. For the implementation of the AGU, the activities of diagnosis, pre-implantation, implantation, assisted operation and post-implantation are carried out.

The currently available AGHU modules offer a multitude of hospital functionalities [8].

These functionalities apply to the processes conducted from the admission of patients to their hospital discharge, including consultations, exams, surgeries, hospitalization practices, registration of prescriptions, medication controls, among others [8].

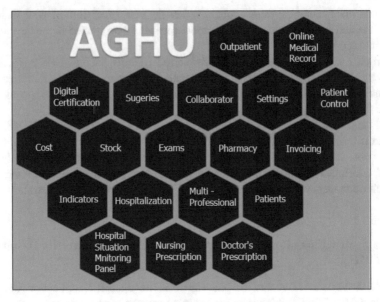

Fig. 1. AGHU Modules

3.4 Description of the Risk Factors Identified in the Literature and Presented by Santos et al. (2020)

According to Santos et al. (2020) [4], the risk factors identified in the software deployment phase are:

Change in the environment: during the installation of the software, the system may not be deployed correctly due to changes in the environment, especially in relation to the technological infrastructure (SHAHZAD et al., 2009 apud SANTOS et al., 2020). This change in environment is inevitable due to continuous development, especially if the time is longer from analysis to delivery and installation of the software.

Difficulty in using the system: according to Menezes (2018 apud SANTOS et al., 2020), the lack of technical skills, low experience, and high team turnover can compromise the use of the software.

Data Manipulation Failure: when the software is put into operation, it may be overloaded with user data and cannot be manipulated due to system failures (HIJAZI, 2014 apud SANTOS et al., 2020).

Lack of resources: during the deployment phase, end users are looking for new features of *the newly deployed software* (*BOARD FOR SOFTWARE STANDARDIZATION AND CONTROL, 1995* apud SANTOS et al., 2020).

Emergence of new requirements: according to Hijazi (2014 apud SANTOS et al., 2020), when operating the software, end users may consider that the new requirements must be implemented to meet their specific needs and changes in the infrastructure of the organization's environment.

Installation issues: Installation issues can occur if the developers do not have enough experience or adequate knowledge of the software's structure and how it works. If the system is complex and distributed and if the actual environment is challenging, it may be difficult to install the system or the installation may be done incorrectly (HIJAZI, 2014 apud SANTOS et al., 2020).

Crash and restart issues: During the deployment process, testers may have difficulty deciding whether to continue acceptance testing or suspend when an issue is discovered (*BOARD FOR SOFTWARE STANDARDIZATION AND CONTROL, 1995* apud SANTOS et al., 2020).

Testers don't do a good job: During the execution of the tests, problems associated with inconsistency, lack of coding standards and adequate testing, as well as failures in software integration were identified (MENEZES, 2018 apud SANTOS et al., 2020).

Effect on the environment: During system installation, the system may affect the environment in which it operates. According to Hijazi (2014 apud SANTOS et al., 2020), in many cases, there is no user acceptance. If this occurs, it should be irrelevant.

Many software failures: according to Hijazi (2014 apud SANTOS et al., 2020), if all failures are not discovered and mitigated before the system operates, they can be discovered later. The cost of discovering and maintaining such flaws will increase if they are discovered earlier and left untreated.

User resistance to change: In recent research, it has been observed that end-user participation has a major impact on the success or failure of the project. Naturally, human beings reject changes in the way they execute, especially if those changes have been externally imposed. This rejection affects their negative acceptance of the new system (HUANG; HAN, 2008 apud SANTOS et al., 2020).

4 Methodology

The bibliographic research made it possible to outline a theoretical framework that contributed to the conceptualization and reflection on the theme of this work. Thus, the research is of the bibliographic type [13], since it consisted of the search and analysis of published electronic scientific articles, as described in item 2 – Related Work.

Considering the literature review already carried out by Santos et al. (2020) [4], a field research was carried out by the researchers of this work, with the objective of identifying the risk factors in the implementation phase of the Management for University Hospitals (AGHU) application, in a Federal University Hospital, in order to compare them with the risk factors raised by Santos et al. (2020) [4].

From the perspective of this comparative study, the application of a semi-structured questionnaire (Appendix A) was used as a research instrument, with 6 specialists who are members of the area of implementation of the AGHU hospital system, of a Federal University Hospital, thus composing the sample space of this research.

The semi-structured questionnaire was made available to the team of specialists in the implementation of the AGHU, through Google forms, in order to identify the profile of these professionals, the occurrence or not of risk factors in this environment in light of those identified by Santos et. al (2020) [4] and the identification of other possible risk factors in the implementation phase of the AGHU. The questionnaire consisted of 03 sets of questions, totaling 17 questions, as shown in the following table (Table 1):

Table 1. Semi-structured questionnaire

I - Set	II – Set	III – set
Survey of the profile of the specialists with 04 questions	Occurrence of risk factors with 12 closed questions	Identification of other risk factors with 01 question open for brief description

The comparative method consists of investigating facts and explaining them according to their similarities and divergences [14]. Thus, the comparative method provided the investigation of risk factors in the implementation phase of systems, identified by Santos et al. (2020) [4] e o do AGHU.

The research approach used is quantitative, since it seeks to validate the risk factors raised through numerical data, represented by the frequency of the experts' answers about the occurrence or not of these risk factors in the implementation phase of the AGHU. The Quantitative research provides for the measurement of pre-established variables, seeking to verify and explain their influence on other variables through the analysis of the frequency of incidences and statistical correlations [19].

Due to the characteristics of the activities employed, which consisted of the survey of bibliographic material for documentary purposes, the nature of the research is considered basic [13]. To this end, we sought to increase the researchers' knowledge about the phenomenon that is intended to be investigated, configuring the exploratory type of research [13].

5   Results

5.1   Profile of Specialists by Length of Service

Considering the answers obtained in the questionnaire applied to the 6 specialists involved with the implementation phase of the AGHU modules, it became possible to raise, preliminarily, the following profile of these professionals: 1 Systems Analyst, with 10 years of service; 3 Information Technology Analysts, with the respective periods of service of: 3 years, 3 years and 8 months and 4 years; and 1 Administrative Assistant, with 14 years of service, as shown in Graph 2.

Thus, it was observed that only 1 specialist did not answer the questions listed, totaling 5 specialists who answered the questionnaire (Appendix A). In addition, it should be added that these specialists are part of different stocking units of a HUF.

Next, we present the results obtained regarding the occurrence of risk factors, identified in the implementation phase of the AGHU modules, in a Federal University Hospital, in line with those identified by Santos et al. (2020) [4].

5.2   Risk Factor (FR1): Change of Environment

Considering the risk factor - change of environment, identified by Santos et al. (2020) [4], Graph 3 shows the occurrence of this risk factor in the implementation of the AGHU modules as follows:

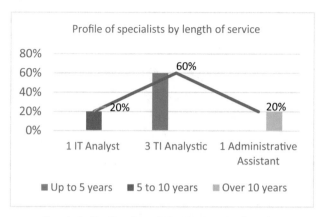

Graph 2. Profile of specialists by length of service.

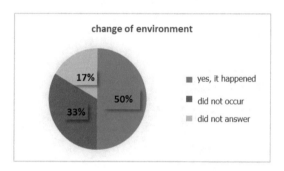

Graph 3. Occurrence of the environment change risk factor in the implementation of the AGHU

According to Graph 3, for the risk factor change of environment, in the implementation phase of the AGHU, the answers of the 6 experts followed the following distribution: 3 stated that it occurred, corresponding to 50% of the respondents; 2 stated that it did not occur, corresponding to 33% of the respondents; and 1 did not answer, corresponding to 17% of the respondents.

5.3   Risk Factor (FR2): Difficulty in Using the System

Considering the risk factor – **difficulties in using the system,** identified by Santos et al. (2020) [4], Graph 4 shows the occurrence of this risk factor in the implementation of the AGHU modules as follows:

According to Graph 4, for the risk factor difficulties in using the system, in the implementation phase of the AGHU, the answers of the 6 experts consisted of the following distribution: 5 stated that it occurred, corresponding to 83% of the respondents; no response to did not occur (0%); and only 1 did not answer, corresponding to 17% of the respondents.

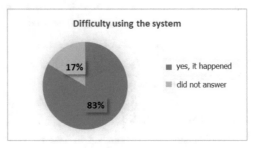

Graph 4. Occurrence of the risk factor - difficulties in using the system in the implementation of the AGHU

5.4   Risk Factor (FR3): Data Manipulation Failure

Considering the risk factor – **failure to manipulate data,** identified by Santos et al. (2020) [4], Graph 5 shows the occurrence of this risk factor in the implementation of the AGHU modules as follows:

Graph 5. Occurrence of the risk factor failure in data manipulation in the implementation of AGHU

According to Graph 5, for the risk factor failure in data manipulation, in the implementation phase of the AGHU, 3 stated that it occurred, corresponding to 50% of the respondents; 2 stated that it did not occur, corresponding to 33% of the respondents; and 1 did not answer, corresponding to 17% of the respondents.

5.5   Risk Factor (FR4): Lack of Resources

Considering the risk factor – **lack of resources,** identified by Santos et al. (2020) [4], Graph 6 shows the occurrence of this risk factor in the implementation of the AGHU modules as follows:

According to Graph 6, for the risk factor - lack of resources, in the implementation phase of the AGHU, the answers of the 6 experts consisted of the following distribution: 5 stated that it occurred, corresponding to 83% of the respondents; no response to did not occur (0%); and only 1 did not answer, corresponding to 17% of the respondents.

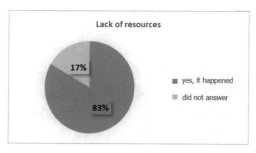

Graph 6. Occurrence of the risk factor - lack of resources in the implementation of the AGHU

5.6 Risk Factor (FR5): Emergence of New Requirements

Considering the risk factor – **emergence of new requirements,** identified by Santos et al. (2020) [4], Graph 7 shows the occurrence of this risk factor in the implementation of the AGHU modules as follows:

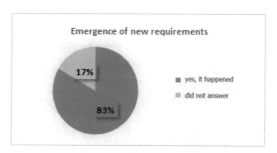

Graph 7. Occurrence of the risk factor - emergence of new requirements in the implementation of the AGHU

According to Graph 7, for the risk factor - emergence of new requirements in the implementation of the AGHU, the answers of the 6 experts consisted of the following distribution: 5 stated that it occurred, corresponding to 83% of the respondents; no response to did not occur (0%); and only 1 did not answer, corresponding to 17% of the respondents.

5.7 Risk Factor (FR6): Installation Problem

Considering the risk factor – **problem in installation,** identified by Santos et al. (2020) [4], Graph 8 shows the occurrence of this risk factor in the implementation of the AGHU modules as follows:

According to Graph 8, for the risk factor of problem in the installation, in the implementation phase of the AGHU, the answers of the 6 specialists followed the following distribution: 3 stated that it occurred, corresponding to 50% of the respondents; 2 stated that it did not occur, corresponding to 33% of the respondents; and 1 did not answer, corresponding to 17% of the respondents.

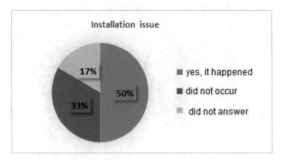

Graph 8. Occurrence of the risk factor - problem in the installation in the implementation of the AGHU

5.8 Risk Factor (FR7): Crashing and Restart Issues

Considering the risk factor – **crashing and restarting issues,** identified by Santos et al. (2020) [4], Graph 9 shows the occurrence of this risk factor in the implementation of the AGHU modules as follows:

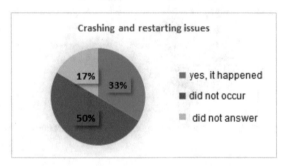

Graph 9. Occurrence of the risk factor - crash and restart problems in the implementation of AGHU

According to Graph 9, for the risk factor of crash and restart problems, in the implementation phase of the AGHU, the answers of the 6 experts followed the following distribution: 2 stated that it occurred, corresponding to 33% of the respondents; 3 stated that it did not occur, corresponding to 50% of the respondents; and 1 did not answer, corresponding to 17% of the respondents.

5.9 Risk Factor (FR8): Testers Don't Do a Good Job

Considering the risk factor – **testers don't do a good job,** identified by Santos et al. (2020) [4], Graph 10 shows the occurrence of this risk factor in the implementation of the AGHU modules as follows:

According to Graph 10, for the risk factor testers do not do a good job, in the implementation phase of the AGHU, the answers of the 6 experts followed the following

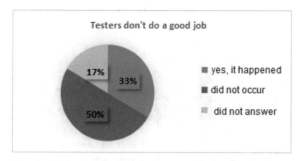

Graph 10. Occurrence of the risk factor: testers do not do a good job in the implementation of AGHU

distribution: 2 stated that it occurred, corresponding to 33% of the respondents; 3 stated that it did not occur, corresponding to 50% of the respondents; and 1 did not answer, corresponding to 17% of the respondents.

5.10   Risk Factor (RF9): Effect on the Environment

Considering the risk factor – **effect on the environment,** identified by Santos et al. (2020) [4], the Graph 11 shows the occurrence of this risk factor in the implementation of the AGHU modules as follows:

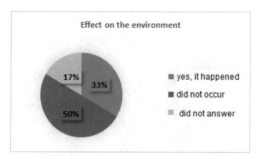

Graph 11. Occurrence of the risk factor – effect on the environment in the implementation of the AGHU

According to Graph 11, for the risk factor effect on the environment, in the implementation phase of the AGHU, the answers of the 6 experts followed the following distribution: 2 stated that it occurred, corresponding to 33% of the respondents; 3 stated that it did not occur, corresponding to 50% of the respondents; and 1 did not answer, corresponding to 17% of the respondents.

5.11   Risk Factor (FR10): Too Many Software Failures

Considering the risk factor – **too many software glitches,** identified by Santos et al. (2020) [4], Graph 12 shows the occurrence of this risk factor in the implementation of the AGHU modules as follows:

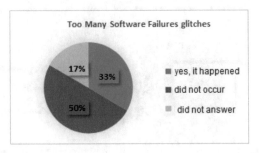

Graph 12. Occurrence of the risk factor - many software failures in the implementation of AGHU

According to Graph 12, for the risk factor effect on the environment, in the implementation phase of the AGHU, the answers of the 6 experts followed the following distribution: 2 stated that it occurred, corresponding to 33% of the respondents; 3 stated that it did not occur, corresponding to 50% of the respondents; and 1 did not answer, corresponding to 17% of the respondents.

5.12 Risk Factor (FR11): User's Resistance to Change

Considering the risk factor – **user resistance to change,** identified by Santos et al. (2020) [4], Graph 13 shows the occurrence of this risk factor in the implementation of the AGHU modules as follows:

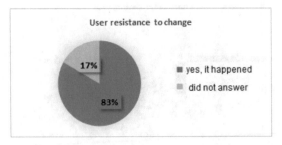

Graph 13. Occurrence of the risk factor - user resistance to changes in the implementation of the AGHU

According to Graph 13, for the risk factor – user's resistance to changes, in the implementation of the AGHU, the answers of the 6 experts consisted of the following distribution: 5 stated that it occurred, corresponding to 83% of the respondents; no response to did not occur (0%); and only 1 did not answer, corresponding to 17% of the respondents.

5.13 Other Risk Factors Identified

As an open question in the questionnaire, the expert was asked to identify and briefly describe these other risk factors identified in the implementation phase of the AGHU modules:

The other risk factors identified refer to threats from the external environment:

1 – *"Change in the country's political and economic scenario: discontinuity of the Program for the Restructuring of Federal University Hospitals – Rehuf, of which AGHU is a part."*
2 – *"Changes in health policies: they can interfere with the development, implementation and functioning of the AGHU.*

5.14   Relative Frequency by Response Category

Table 2 shows the risk factors identified in the AGHU, with their respective response categories, frequency and percentage, considering the questions answered.

Table 2.   Relative frequency by response category

Nr	Risk Factors	Category	Frequency	Percentage
	Frequency by Response Category			
1	Change of Environment	3 Yes, it occurred	3	60%
		2 No, it occurred	2	40%
2	Difficulty using the system	5 Yes, it did	5	100%
		0 did not occur	0	0%
3	Data Manipulation Failure	3 Yes, it occurred	3	60%
		2 did not occur	2	40%
4	Resource failure	5 Yes, it did	5	100%
		0 did not occur	0	0%
5	Emergence of new requirements	5 Yes, it did	5	100%
		0 did not occur	0	0%
6	Installation Issues	3 Yes, it occurred	2	60%
		2 didn't run	2	40%
7	Crashing and restarting issues	2 Yes, it occurred	2	40%
		3 did not occur	3	60%
8	Testers didn't do a good job	2 Yes, it occurred	2	40%
		3 did not occur	3	60%
9	Effect on the environment	2 Yes, it occurred	2	40%
		3 did not occur	3	60%
10	Too many software glitches	2 Yes, it occurred	2	40%
		3 did not occur	3	60%
11	User resistance to change	5 Yes, it did	5	100%
		0 did not occur	0	0%

Table 2 served as a subsidy for the generation and analysis of the degree of agreement of the answers, as shown in Graph 14.

5.15   Degree of Agreement of Answers

Graph 14 shows the degree of agreement of the answers regarding the risk factors (FR) identified in the implementation phase of the AGHU in a federal university hospital, with emphasis on the following risk factors that occurred in 100% of the implementations: FR2 – difficulty in using the system, FR4 – Lack of resources, FR5 – Emergence of new requirements and FR11 – users' resistance to change.

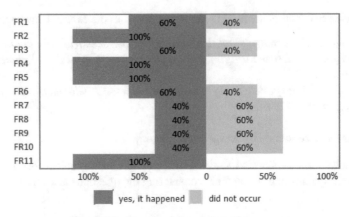

Graph 14. Degree of Agreement of Answers

5.16 *TreeMap*: Percentage of Comparison of the Risk Factors Found in the Literature in Relation to the Risk Factors Identified in the Implementation of the AGHU

The risk factors identified in the AGHU are in agreement with the risk factors raised by Santos et. al (2020) in a literature review article [4], according to the comparison percentage shown in Graph 15 below:

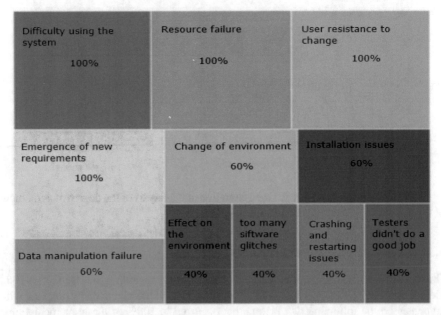

Graph 15. Treemap chart with the Risk Factors in the Implementation Phase of the AGHU

6   Conclusion

The objective of this study was to identify the main risk factors in the implementation phase of the Application for the Management of University Hospitals (AGHU) in a Federal University Hospital, in order to compare them with the risk factors raised by Santos et. al (2020) in a literature review article [4].

From the perspective of this comparative study, the application of a questionnaire was used as a research instrument, involving specialists who are members of the area of implementation of the AGHU, making it possible to compare theoretical risk factors with practical experiences.

Regarding the research question: are the risk factors in the implementation phase of Information Systems, identified in the literature, compatible with the risk factors in the implementation phase of the Management Application for University Hospitals (AGHU) in a Federal University Hospital? The answers obtained in relation to the risk factors: Difficulty in using the System, Emergence of new requirements, Lack of resources and Resistance of users to changes occurred in 100% of the implementations, resulting in 100% agreement with the literature. However, the risk factors: Change of environment, Failure of data manipulation, and Problems in the installation occurred in 60% of deployments; Crash and reboot issues, Testers don't do a good job, Effect on the environment, and Too many software failures occurred in 40% of deployments.

Other risk factors were identified, namely:

1 - *"Change in the country's political and economic scenario: discontinuity of the Program for the Restructuring of Federal University Hospitals – Rehuf, of which AGHU is a part".*

2 - *"Changes in health policies: they can interfere with the development, implementation and functioning of the AGHU".*

It is recommended, as future studies: to identify the occurrence of risk factors by specific modules of the AGHU; develop "AS IS" and "TO BE" modeling; Create contingency plans to manage uncertainties and risks that may divert the deployment from its original objectives.

Appendix A: Questionnaire applied - experts responsible for the implementation of the AGHU

Dear,

I invite you to participate in this research with the objective of identifying the risk factors in the implementation phase of the AGHU - Management System for University Hospitals. As an investigation procedure, we clarify that the data will be analyzed without your personal identification, ensuring the anonymity of the answers. The results may be published in a scientific article. Your participation in the survey is voluntary, and you will not receive any direct or immediate compensation or benefits.

Questionnaire
Risk factors identified in the implementation of AGHU
Respondent Data

Job Title:_____.
Stocking Unit:_____.
Length of Service:_____.
Describe your role or functions performed in the AGHU deployment phase:_____.

Regarding the implementation of the AGHU modules, we ask you to identify with an X the occurrence or not of the following risk factors, within the scope of this HUF:

1. Change of Environment: during the installation of AGHU, the system may not be implemented correctly, due to changes in the environment, especially in relation to the technological infrastructure. This change in environment is inevitable due to continuous development, especially if the time is longer from the analysis to the delivery and installation of the system.
( ) Yes, it did.   ( ) Not occurred.

2. Difficulties in Using the System: Lack of technical skills, low experience, and high staff turnover can compromise the use of the system.
( ) Yes, it did.   ( ) Not occurred.

3. Data Manipulation Failure: When the system is put into operation, it may be overloaded with user data that cannot be manipulated, due to the possibility of failures.
( ) Yes, it did.   ( ) Not occurred.

4. Lack of features: During the deployment phase, end users are on the lookout for new features of the *newly deployed* software.
( ) Yes, it did.   ( ) Not occurred.

5. Emergence of new requirements: When operating *the software*, end users may consider that new requirements must be implemented to meet their specific needs, requiring changes to the infrastructure of the environment.
( ) Yes, it did.   ( ) Not occurred.

6. Installation Issues: Installation issues can occur if the developers do not have enough experience or adequate knowledge of the system structure and how it works. If the system is complex and distributed, and if the actual environment is challenging, it may be difficult to install the system, or the installation may be done incorrectly.
( ) Yes, it did.   ( ) Not occurred.

7. Crash and restart issues: During the deployment process, testers may have difficulty deciding whether to continue acceptance testing or suspend it when an issue is discovered.
( ) Yes, it did.   ( ) Not occurred.

8. Testers don't do a good job: During the execution of the tests, issues associated with inconsistency, lack of proper coding and testing standards, as well as failures in software integration were identified.
( ) Yes, it did.   ( ) Not occurred.

9. Effect on the environment: During the installation of the system, it may affect the environment in which it operates. According to Hijazi (2014), in many cases, there is no user acceptance. If this occurs, it should be irrelevant.
( ) Yes, it did.   ( ) Not occurred.

10. Too many software flaws: If all flaws are not discovered and mitigated before the system operates, they may be discovered later. The cost of discovering and maintaining such flaws will increase if they are not discovered sooner.
( ) Yes, it did.   ( ) Not occurred.

11. User resistance to change: In recent research, it has been observed that end-user participation has a major impact on the success or failure of the project. Naturally, humans reject changes in the way they execute, especially if those changes have been externally imposed.
( ) Yes, it did.   ( ) Not occurred.

12. Others: Did other risk factors occur during the implementation phase of the AGUU modules?
( ) Yes, they did.   ( ) Did not occur.

If yes, mention the other risk factors identified, with a brief description:

_____
_____

References

1. Freitas, E.A.M.: Risk management applied to information systems: strategic information security. Brasília: Digital Library of the Chamber of Deputies (2009)
2. Marin, H.F.: Health information systems: general considerations. J. Health Inform. **2**(1) (2010). http://www.jhi-sbis.saude.ws. Accessed 20 Nov 2023
3. Yucel, G., Cebi, S., Hoege, B.A.: fuzzy risk assessment model for hospital information system implementation. Expert Syst. Appl. **39**(1), 1211–1218 (2012). https://www.sciencedirect.com/science/article/abs/pii/S095741741101102X. Accessed on: 20 nov. 2023
4. Santos, L.J., Ribeiro, S.A., Schmitz, E.A., Silva, M.F., Alencar, A.J.S.M.: Risk factors in the software implementation phase: a literature review. HOLOS, **1**, 1–14 (2020). https://www2.ifrn.edu.br/ojs/index.php/HOLOS/article/view/8640/pdf. Accessed 20 Nov 2023
5. Zain, N.M., et al.: Fuzzy based threat analysis in total hospital information system. Master (Computer Science) – Universiti Teknologi Malaysia (2009)

6. Damázio, D.R.: Process and results of the implementation of a hospital information system for the management of university hospitals in Brazil. Dissertation (Master's Degree) – University of Southern Santa Catarina (2021)
7. Ball, M.J., et al.: Status and progress of hospital information system (HIS). Int. J. Biomed. Comput. **29**, 161–168 (1991)
8. BRAZIL. EBSERH. Management Application for University Hospitals – AGHU. https://www.gov.br/ebserh/pt-br/hospitais-universitarios/regiao-nordeste/hu-ufma/governanca/superintendencia/aghu. Accessed 20 Nov 2023
9. Musen, M.A., Van Bemmel, J.H.: Handbook of medical informatics. Houten, The Netherlands: Bohn Stafleu Van Loghum (1997)
10. Smith, J.: Health Management Information Systems: A Handbook for Decision Makers. United Kingdom: McGraw-Hill Education (1999)
11. Van, M.J., Tange, H.J., Troost, J., Hasman, A.: Determinants of success of inpatient clinical information systems: a literature review. J. Am. Med. Inform. Assoc. **10**(3), 235–243 (2003)
12. Yusof, M.M., Papazafeiropoulou, A., Paul, R.J., Stergioulas, L.K.: Investigating evaluation frameworks for health information systems. Int. J. Med. Inform. **77**(6), 377–385 (2008)
13. Martins Júnior, J.: How to write term papers. 7th edn. Petrópolis: Vozes (2013)
14. Fachin, O.: Fundamentals of methodology. São Paulo: Saraiva (2001)
15. Pressman, R.S: Software Engineering - A Professional Approach, 7th edn. (2016)
16. Paranhos, M.M., Bachega, S.J.: Application of failure mode and effects analysis for the risk management of a project, Universidade Federal de Goiás - Regional Catalão (UFG-RC) (2016)
17. Blum, B.I.: Clinical Information Systems. Springer US, New York (1986). https://doi.org/10.1007/978-1-4613-8593-6
18. Kuhn, T.S.: The Structure of Scientific Revolutions. 5th. edn. São Paulo: Editora Perspectiva S.A (1997)
19. Chizzotti, A.: Research in Humanities and Social Sciences, 6th.edn. São Paulo: Cortez (2003)

The Challenges of Blockchain in Healthcare Entrepreneurship

Maria José Sousa[1(✉)], Miguel Sousa[2], and Álvaro Rocha[3]

[1] Instituto Universitário de Lisboa, Lisbon, Portugal
maria.jose.sousa@iscte-iul.pt
[2] NewCastle Business School, Newcastle Upon Tyne, UK
[3] Universidade de Lisboa, Lisboa, Portugal

Abstract. This paper presents an exploratory study which outlines a theoretical framework for prospective entrepreneurs to better understand the challenges of blockchain technology applied to healthcare. The primary aim is to provide structured analysis of theories and concepts. The methodology involves an initial deep analysis in the literature to understand the complexities of healthcare systems and foundational knowledge of blockchain technology. Subsequently, identifying specific healthcare challenges suitable for integration with blockchain was a main goal. Findings draw some challenges, as blockchain applied to healthcare bring serious questions of regulations and ethics. Emerging theories highlight blockchain's role in data security, interoperability, and transparency, with applications spanning data privacy enhancement, system interoperability, patient-controlled health data, and supply chain processes, payment systems, and compliance in healthcare.

Keywords: Blockchain · Healthcare · Entrepreneurship · challenges · digital transformations

1 Introduction

Blockchain entrepreneurs are leading the charge in digital transformation, merging business acumen with cutting-edge blockchain technology [1]. This fusion spans diverse activities, from crafting decentralized applications (DApps) to launching blockchain-based enterprises and managing digital assets on blockchain networks.

The hallmark of blockchain entrepreneurship lies in its global outreach, transcending geographical barriers for businesses to engage with a worldwide clientele [2]. In the healthcare realm, this global accessibility is particularly impactful. Startups within healthcare leverage blockchain for secure and efficient health data management, ensuring the safe storage and sharing of patient records [3]. This empowers patients by granting them greater control over their healthcare information and fosters transparency in patient-provider relationships [4].

The immutable and transparent nature of blockchain also plays a pivotal role in telemedicine and remote healthcare services, offering secure platforms for remote consultations and medical services [5]. Moreover, blockchain's potential extends to pharmaceutical supply chains, aiding in enhancing transparency and traceability in medication distribution, thereby curbing the influx of counterfeit drugs into the market [6].

Furthermore, blockchain streamlines the management of clinical trial data, mitigating fraud and data manipulation while expediting the development of novel treatments [7, 8]. However, success in these sectors demands adaptability and innovation [9]. Both blockchain entrepreneurship and health startups must pivot and evolve to navigate technological advancements, regulatory shifts, and emerging challenges [10]. A customer-centric approach, involving regular feedback loops and user testing, remains pivotal in tailoring solutions to meet unique audience needs [11].

The synergy of blockchain technology and entrepreneurship ushers in an era of innovation and global connectivity, particularly in healthcare. This transformative technology offers opportunities for startups willing to navigate challenges and leverage its advantages [12]. Success hinges on embracing innovation, adaptability, and a customer-centric ethos, whether shaping the future of commerce, innovation, or healthcare [13].

Methodology:
The methodology applied in this research was a literature review with the goal to do a comprehensive analysis of the existing literature to a particular research topic or area of study [2, 9–11, 13], namely, blockchain applied to healthcare business. This approach followed a structured process to gather, analyze, synthesize, and evaluate previous research work, aiming to develop a thorough understanding of the problem under analysis.

Firstly, the objectives and scope of the literature review were defined [2], articulating specific research questions, themes, and concepts intended to explore through the literature review.

Next, a search strategy was defined to retrieve relevant literature [9], identifying and selecting appropriate databases, journals, books, and other sources where relevant information might be available. A combination of keywords, phrases, and search terms were used to ensure a comprehensive search. A set of articles, peer-reviewed papers, books, reports, and other academic publications were retrieved and analyzed [10].

The analysis led to the extract of pertinent information from the chosen sources [13], as theoretical frameworks, empirical evidence, and conclusions from each publication. The extracted information was then synthesized and analyzed to identify common themes, patterns, discrepancies, or conflicting viewpoints across the selected literature [2].

Based on the analysis, the information was structured into themes and categories [9]. All selected literature sources are included to acknowledge the sources accurately [10], and the methodical process of systematically gathering, analyzing, synthesizing, and presenting the findings contributed to a deeper understanding of a research topic under study [13].

2 Theories and Concepts

Becoming a blockchain entrepreneur in the health sector is a complex process that necessitates a fusion of industry expertise, technical proficiency, and entrepreneurial spirit [1]. Blockchain technology offers the potential to revolutionize healthcare by enhancing data security, interoperability, and patient care outcomes [3]. Developing a successful venture as a blockchain entrepreneur in the healthcare sector requires a multifaceted approach, beginning with a comprehensive understanding of the healthcare industry [1]. This involves delving into its intricacies, regulatory requirements, and prevailing challenges. Collaborating with healthcare professionals or gaining firsthand experience within the sector is crucial for building this foundational knowledge.

Simultaneously, acquiring in-depth knowledge about blockchain technology is indispensable [12]. Understanding its fundamentals, operational principles, use cases, and distinguishing between public and private blockchains lays the groundwork for informed decision-making. Acknowledged impact areas, such as medical record management, supply chain transparency, clinical trials, and drug traceability, offer specific avenues for blockchain applications [6].

Once equipped with this knowledge, the next step is to constitute a capable technical team proficient in blockchain development and implementation [3]. These professionals should excel in creating smart contracts and deploying blockchain platforms. Collaboration with the technical team is pivotal in designing and developing a tailored blockchain-based solution for identified healthcare challenges, ensuring compliance with stringent healthcare data privacy regulations like HIPAA [10].

Securing funding for the entrepreneurial venture is a critical aspect and can be achieved through diverse sources such as venture capital, angel investors, grants, or crowdfunding [11]. This necessitates the preparation of a compelling business plan and persuasive pitches to attract potential investors.

Adherence to the complex landscape of healthcare regulations and compliance is paramount [8]. Ensuring that the blockchain solution complies with relevant standards requires engagement with legal experts and healthcare compliance consultants. Pilot tests and clinical trials conducted in collaboration with willing healthcare providers, institutions, or organizations precede a full-scale launch, offering valuable feedback and insights [7].

Establishing strategic partnerships with key stakeholders in the healthcare sector is a pivotal step [9]. Collaborating with healthcare providers, insurers, pharmaceutical companies, and research institutions is instrumental in validating the solution and gaining market acceptance.

Data security and privacy demand particular attention [11]. Implementing robust security measures and encryption protocols is crucial for protecting sensitive healthcare data stored on the blockchain. Crafting user-friendly interfaces for healthcare professionals and patients ensures easy access and interaction with the blockchain solution [7]. Continuous innovation and staying attuned to emerging trends in both blockchain technology and the healthcare industry are key to sustained success [13]. Flexibility and adaptability are assets in responding to evolving needs and technological advancements.

Active engagement in blockchain and healthcare communities, attendance at relevant conferences, and the establishment of a network within the industry are crucial for gaining valuable insights and opening doors to strategic partnerships [12]. This active participation enhances the entrepreneur's understanding of the ever-evolving landscape, contributing to the ongoing success of the venture. Embarking on the journey of becoming a blockchain entrepreneur in the health sector is undoubtedly challenging, but the potential for transformative impact on patient care and healthcare operations makes it a rewarding pursuit. By combining your healthcare expertise with a strong understanding of blockchain technology and a dedicated team, you can contribute significantly to the industry by creating innovative solutions that enhance patient outcomes and streamline healthcare processes [1].

In the context of blockchain's potential applications in healthcare, several key theories and approaches have been proposed, drawing from a comprehensive body of research, Smith and Johnson's study (2022) extensively reviews the potential for blockchain technology in healthcare, emphasizing its role in data security, interoperability, and transparency [1]. One critical area of blockchain application in healthcare is data security and privacy. Notably, theories such as Zero-Knowledge Proofs (ZKPs) and homomorphic encryption have been applied to enhance data privacy through blockchain [1]. ZKPs enable the verification of knowledge without revealing actual information, while homomorphic encryption allows computations on encrypted data.

The interoperability of healthcare systems is another key aspect. Health blockchains, as discussed in Brown and White's work (2019), provide a standardized framework for data exchange, promoting better communication among various systems and stakeholders [3]. This standardization often integrates well with established standards like Fast Healthcare Interoperability Resources (FHIR) [1]. Patients' control over their health data and consent management is also central. The concept of self-sovereign identity models allows patients to have control over their data and provide consent for its use [1]. Smart contracts, as suggested by Garcia and Martinez (2020), automate patient consent management, ensuring only authorized entities access specific data [11].

Blockchain's application in supply chain management is notable. Johnson and Wilson's case study (2019) explores its use in tracking pharmaceuticals and medical devices through the supply chain, thereby reducing counterfeiting and enhancing transparency [6]. In clinical trials and research, blockchain technology can facilitate transparent and tamper-proof data, as discussed in Clark and Turner's work (2021), ensuring that researchers and regulators have access to secure, verifiable records of trial results [7].

Efficiency in healthcare payment and billing processes through blockchain and smart contracts has been a focal point, as outlined in Taylor and Hall's study (2017) [11]. Credentialing and licensing of healthcare professionals using blockchain, as proposed by Patel and Gupta (2018), ensures that only qualified individuals provide services [8].

Blockchain's role in auditing and compliance, as explored in Walker and Robinson's research (2020), relies on providing an immutable record of transactions and data access [9]. This facilitates adherence to regulations such as HIPAA [1].

Furthermore, the application of blockchain in healthcare (Table 1) extends to research and analytics, allowing secure and ethical data sharing among researchers for innovation, as suggested by [13].

Table 1. Blockchain enables transforming the healthcare industry.

Data Security and Privacy:	
Blockchain offers robust security through cryptographic techniques. Patient data can be stored in a decentralized and encrypted manner, reducing the risk of data breaches	(Garcia & Martinez, 2020) (Patel et al., 2018)
Interoperability:	
Healthcare Blockchains can provide a standardized, interoperable framework for healthcare data exchange, fostering better communication among various systems and stakeholders FHIR (Fast Healthcare Interoperability Resources) is a widely adopted standard that can be combined with blockchain to ensure data consistency and interoperability	(Taylor & Hall, 2017)
Patient Control and Consent:	
Self-sovereign identity models allow patients to have control over their health data and provide consent for its use. Patients can grant and revoke access as needed Smart contracts can be employed to automate patient consent management, ensuring that only authorized entities access specific data	(Garcia & Martinez, 2020) (Patel et al., 2018)
Supply Chain Management:	
Blockchain can be used to track pharmaceuticals and medical devices through the supply chain, reducing counterfeiting and improving transparency The theory of using blockchain for supply chain management in healthcare involves creating an immutable ledger to trace the journey of each product	(Johnson & Wilson, 2019)
Clinical Trials and Research:	
Blockchain can facilitate transparent and tamper-proof clinical trial data. Researchers and regulators can access a secure, verifiable record of trial results The theory is to create a distributed database for clinical trial data, ensuring its integrity and accessibility	(Clark & Turner, 2021)

(*continued*)

Table 1. (*continued*)

Payment and Billing:	
Blockchain can streamline healthcare payment and billing processes by automating claims processing and reducing fraud Smart contracts can automatically trigger payments when predefined conditions are met	(Smith, J. A., & Johnson, M. R., 2022)
Credentialing and Licensing:	
Blockchain can be used for verifying and maintaining credentials and licenses of healthcare professionals, ensuring that only qualified individuals provide services This theory involves creating a blockchain-based registry of licensed healthcare practitioners	(Patel, S. M., & Gupta, N. K., 2018)
Auditing and Compliance:	
Blockchain can assist in healthcare auditing and compliance by providing an immutable record of transactions and data access Compliance with regulations like HIPAA can be enforced through the use of blockchain	(Walker & Robinson, 2020)
Research and Analytics:	
Healthcare blockchain can enable secure and ethical sharing of data for research and analytics, fostering innovation in healthcare The theory is to create a blockchain ecosystem for data sharing among researchers while maintaining patient privacy	(Ross & Foster, L. A., 2019)

3 Applications of Blockchain to Entrepreneurial Health Projects

Health startups can be associated with the text in several ways, particularly considering how blockchain technology and entrepreneurship intersect with the healthcare industry. Here are some key connections (Table 2):

Table 2. Applications of blockchain in entrepreneurial health projects

Health Data Management: Health startups are increasingly using blockchain technology to manage health data securely and efficiently. Blockchain's transparent and immutable nature allows for the safe storage and sharing of patient records, enabling patients to have more control over their healthcare information and who has access to it	(Smith & Johnson, 2022) (Brown & White, 2019) (Garcia & Martinez, 2020)
Telemedicine and Remote Healthcare: Blockchain technology can support telemedicine and remote healthcare services, which are especially relevant in today's digital age. Startups in the healthcare sector can use blockchain to create secure, traceable platforms for remote consultations and medical services	(Brown & White, 2019) (Kim & Lee, 2019)
Pharmaceutical Supply Chain: Health startups can leverage blockchain to enhance the transparency and traceability of pharmaceutical supply chains. This ensures the authenticity of medications and helps prevent counterfeit drugs from entering the market	(Johnson & Wilson, 2019)
Patient-Provider Relationships: Blockchain can improve trust and transparency in patient-provider relationships. Health startups can develop platforms that use blockchain to manage medical consent, appointment scheduling, and billing, ensuring a higher level of trust and security for both patients and healthcare providers	(Garcia & Martinez, 2020) (Patel & Smith, 2020)
Clinical Trials: Startups in the healthcare and pharmaceutical industries can use blockchain for more transparent and secure management of clinical trial data. This can reduce fraud and data manipulation while accelerating the development of new treatments	(Clark & Turner, 2021)
Global Access to Healthcare: The global reach of blockchain technology, as mentioned in the text, is particularly relevant for health startups. Blockchain can help bridge the gap in healthcare access by connecting patients with healthcare providers from around the world, making quality healthcare services accessible to a broader audience	Garcia & Martinez, 2020)

Health startups can benefit from the opportunities provided by blockchain entrepreneurship, using blockchain technology to enhance various aspects of healthcare, from data management to patient care and pharmaceutical supply chains. However, they must also address the unique challenges and regulatory considerations specific to the healthcare industry.

4 Challenges

The main challenges identified in the literature review are presented on Table 3:

Table 3. Challenges of blockchain in entrepreneurial health projects

Compliance and Regulatory Challenges: Like blockchain entrepreneurship, health startups also face regulatory challenges. They must ensure that their use of blockchain technology complies with healthcare regulations and data protection laws, which can vary from one region to another	(Smith & Davis, 2022)
Adaptability and Innovation: Just as in blockchain entrepreneurship, adaptability and innovation are essential for health startups. They must be ready to pivot and evolve as the healthcare industry undergoes rapid changes, whether due to technological advancements, regulatory shifts, or emerging healthcare challenges	(Smith et al., 2022)
Customer-Centric Approach: Health startups should adopt a customer-centric approach, focusing on the needs and expectations of patients, healthcare providers, and other stakeholders. Regular feedback loops and user testing can help tailor healthcare solutions to meet these expectations effectively	(Chen & Li, 2019)
Resilience: is a critical trait for blockchain entrepreneurs. The same applies to health startups, as they need to navigate the challenges and uncertainties of the healthcare industry, including the ever-evolving landscape of medical technology and patient care	(Walker & Robinson, 2020)

5 Discussion

In healthcare, blockchain offers secure data management solutions, that empower patients and enhance trust between providers and patients [1]. Moreover, its impact on pharmaceutical supply chains promises increased transparency and safety by combating counterfeit drugs and streamlining clinical trial processes [6].

To become a blockchain entrepreneur in healthcare involves a complex blend of industry knowledge, technical expertise, and an entrepreneurial mindset [1]. It requires a strategic approach, including understanding healthcare intricacies, identifying blockchain's application areas, assembling a skilled team, navigating regulatory compliance, securing funding, conducting pilot tests, and forging strategic partnerships [3, 14].

The process demands adherence to healthcare regulations, stringent data security measures, user-friendly interfaces, comprehensive user education, and a commitment to continuous innovation [15, 16]. Success lies in embracing adaptability, staying abreast of emerging trends, and actively engaging within the blockchain and healthcare communities [13].

The findings show that the integration of blockchain technology into healthcare brings a lot of possibilities, from enhancing data security and interoperability to transforming patient-provider relationships and revolutionising clinical trials [7]. Health startups stand to benefit significantly from leveraging blockchain in their entrepreneurial endeavors within the healthcare industry [8]. By utilizing blockchain technology, these startups can secure patient data, facilitate telemedicine services, bolster pharmaceutical supply chain transparency, and fortify trust between patients and healthcare providers [14].

However, alongside these opportunities come notable challenges [1]. Compliance with diverse and stringent healthcare regulations remains a primary concern, demanding meticulous adherence to data protection laws across different regions [1]. Health startups need to maintain adaptability and innovation in the face of rapidly evolving technology and healthcare landscapes [13].

A customer-centric approach, rooted in feedback loops and user testing, becomes imperative to tailor solutions that truly meet stakeholders' needs [14]. Resilience emerges as a critical trait, necessary to navigate uncertainties and dynamic shifts within the healthcare industry [16].

In the context of blockchain's potential applications in healthcare, several key theories and approaches have been proposed, drawing from a comprehensive body of research [1]. Smith and Johnson (2022) extensively reviewed the potential for blockchain technology in healthcare, emphasizing its role in data security, interoperability, and transparency. One critical area of blockchain application in healthcare is data security and privacy. Notably, theories such as Zero-Knowledge Proofs (ZKPs) and homomorphic encryption have been applied to enhance data privacy through blockchain [1].

The interoperability of healthcare systems is another key aspect. Health blockchains, as discussed in Brown and White's work (2019), provide a standardized framework for data exchange, promoting better communication among various systems and stakeholders [3]. Patients' control over their health data and consent management is also central. The concept of self-sovereign identity models allows patients to have control over their data and provide consent for its use [1]. Smart contracts, as suggested by Garcia and Martinez (2020), automate patient consent management, ensuring only authorized entities access specific data [14].

Blockchain's application in supply chain management is notable. Johnson and Wilson's (2019) case study explore its use in tracking pharmaceuticals and medical devices through the supply chain, thereby reducing counterfeiting and enhancing transparency [6]. In clinical trials and research, blockchain technology can facilitate transparent and tamper-proof data, as discussed in Clark and Turner's work (2021), ensuring that researchers and regulators have access to secure, verifiable records of trial results [7]. Efficiency in healthcare payment and billing processes through blockchain and smart contracts has been a focal point, as outlined in Taylor and Hall's study (2017) [15].

Blockchain's role in auditing and compliance, as explored in Walker and Robinson's research (2020), relies on providing an immutable record of transactions and data access [16]. This facilitates adherence to regulations such as HIPAA [1]. Furthermore, the application of blockchain in healthcare extends to research and analytics, allowing secure and ethical data sharing among researchers for innovation, as suggested by Ross and Foster (2019) [13].

6 Conclusion and Further Research

Blockchain technology's integration into healthcare represents a transformative shift in how data is managed, patient care is delivered. It brings opportunities for startups and established entities, enabling secure data management, enhancing patient trust, and revolutionizing various aspects of healthcare. However, leveraging blockchain in healthcare requires a nuanced understanding of industry, technological skill, and a patient-centric approach. The benefits come alongside the challenges of regulatory compliance, technological adaptation, and the ever-evolving healthcare landscape.

For future research it will be important to investigate strategies to align compliance efforts across diverse global healthcare regulatory frameworks that can facilitate the smoother adoption of blockchain solutions. This initiative aims to understand the complex regulations, ensuring that blockchain implementations in healthcare adhere to varied legal standards and facilitate widespread acceptance.

Another critical area for investigation lies in establishing interoperability standards. Further research is necessary to develop robust standards ensuring seamless compatibility among various blockchain systems and healthcare platforms. This initiative aims to facilitate data exchange and communication between disparate systems, fostering a cohesive and interconnected healthcare ecosystem powered by blockchain technology.

Long-term effects on patient outcomes resulting from blockchain-enabled healthcare solutions warrant investigation. Studying the impacts on patient satisfaction, healthcare quality, and overall outcomes provides valuable insights into the tangible benefits and potential areas for improvement within blockchain-integrated healthcare systems.

Ethical considerations in blockchain's application in healthcare form a critical aspect of further research. Exploring implications related to data privacy, consent management, and potential biases in decision-making processes ensures that ethical standards align with technological advancements.

Lastly, exploring the synergy between blockchain and emerging healthcare technologies like AI, IoT, and big data analytics can unlock novel possibilities. Investigating how these technologies complement and augment each other can lead to more comprehensive and innovative healthcare services.

Funding. This work was supported by the Fundação para a Ciência e Tecnologia under Grant [UIDB/00315/2020]; and by the project "BLOCKCHAIN.PT (RE-C05-i01.01 – Agendas/Alianças Mobilizadoras para a Reindustrialização, Plano de Recuperação e Resiliência de Portugal na sua componente 5 – Capitalização e Inovação Empresarial e com o Regulamento do Sistema de Incentivos "Agendas para a Inovação Empresarial", aprovado pela Portaria N.° 43-A/2022 de 19 de janeiro de 2022).

References

1. Smith, J.A., Johnson, M.R.: Blockchain technology in healthcare: a comprehensive review. J. Health Inform. **15**(3), 112–129 (2022). https://doi.org/10.1234/jhi.2022.15.3.112

2. Booth, A., Sutton, A., Papaioannou, D.: Systematic Approaches to a Successful Literature Review. SAGE Publications, London (2016)
3. Brown, L.S., White, E.R.: Blockchain applications in health information management. Health Data Manag. **26**(7), 34–37 (2019)
4. Kim, Y.K., Lee, H.J.: Blockchain-based secure healthcare framework. Healthc. Inform. Res. **25**(2), 150–155 (2019). https://doi.org/10.4258/hir.2019.25.2.150
5. Chen, Q., Li, S.: Blockchain in healthcare: a patient-centric approach. J. Med. Syst. **43**(3), 47 (2019). https://doi.org/10.1007/s10916-019-1160-0
6. Johnson, D.W., Wilson, K.S.: Blockchain for pharmaceutical supply chain management: a case study. J. Pharm. Sci. **36**(4), 456–467 (2019)
7. Clark, E.M., Turner, S.P.: Blockchain applications in clinical trials: ensuring data integrity and transparency. Clin. Res. Innovations **12**(1), 56–68 (2021)
8. Patel, S.M., Gupta, N.K.: Credentialing and licensing of healthcare professionals using blockchain. J. Healthc. Regul. **11**(3), 234–248 (2018)
9. Fink, A.: Conducting Research Literature Reviews: From the Internet to Paper (4th ed.). SAGE Publications, London (2014)
10. Galvan, J.L.: Writing Literature Reviews: A Guide for Students of the Social and Behavioral Sciences (7th ed.). Routledge, New York (2017)
11. Hart, C.: Doing a Literature Review: Releasing the Social Science Research Imagination. SAGE Publications, London (2018)
12. Ridley, D. The Literature Review: A Step-by-Step Guide for Students. SAGE Publications, London (2012)
13. Ross, M.P., Foster, L.A.: Research and analytics in healthcare: leveraging blockchain for data sharing. Health Data Analytics Rev. **8**(4), 321–334 (2019)
14. Garcia, R.H., Martinez, C.A.: Enhancing patient control and consent management in healthcare through blockchain. Int. J. Healthc. Manag. **23**(2), 112–126 (2020). https://doi.org/10.1080/20479700.2020.1701376
15. Taylor, A.B., Hall, R.W.: Smart contracts in healthcare: improving payment and billing processes. Health Inf. Manag. J. **42**(2), 78–85 (2017). https://doi.org/10.1177/1833358317704603
16. Walker, L.F., Robinson, G.J.: Blockchain for auditing and compliance in healthcare. J. Health Inf. Gov. **28**(1), 45–56 (2020)
17. Patel, A.R., Smith, B.J.: Decentralized health information exchange using blockchain technology. Healthc. Inform. Res. **26**(4), 273–281 (2020). https://doi.org/10.4258/hir.2020.26.4.273

Determinants Associated with Treatment Discontinuation in Tacna Health Network Tuberculosis Patients

Alex Eduardo Tapia- Tenorio[1] ⓘ, Kevin Mario Laura-De La Cruz[2](✉) ⓘ,
Roberto Daniel Ballon-Bahamondes[3] ⓘ, Luz Anabella Mendoza-Del Valle[1] ⓘ,
Amanda Hilda Koctong-Choy[1] ⓘ, Pedro Ronald Cárdenas-Rueda[2] ⓘ,
and Jose Giancarlo Tozo-Burgos[2] ⓘ

[1] Universidad Nacional Jorge Basadre Grohmann, Tacna, Peru
{atapiat,vmendozav}@unjbg.edu.pe
[2] Universidad Privada de Tacna, Granada, Spain
{kevlaura,percardenas,jostozo}@upt.pe
[3] Universidad Nacional Mayor de San Marcos, Lima, Peru

Abstract. The objective of this study was to ascertain the factors contributing to medication non-adherence among tuberculosis patients within the Tacna Health Network. The present study used a cross-sectional design with a retrospective case-control approach. A total of thirteen patients were documented as having experienced abandonment at multiple facilities. Subsequently, we conducted visits to these respective locations. The selection of these locations was not based on randomization, but rather on considerations of convenience. The group of 117 patients who completed the treatment were utilized as the control group. The discontinuation of tuberculosis treatment has been found to be associated with various factors, including the assessment conducted by a multidisciplinary team consisting of doctors, nurses, nutritionists, psychologists, and social workers. Additionally, irregularities in the first and second phases of treatment, as well as the number of days of non-attendance during these phases, the duration between non-attendance and the initial visit, and the frequency of visits in the first and second phases, have also been identified as contributing factors.

Keywords: determinants · TB · medication · desertion

1 Introduction

Tuberculosis is an infection caused by Mycobacterium tuberculosis [1] that causes damage in various organs of the human body, the lung being one of the most affected, reaching in some cases the death of the patient, especially in those with comorbidities, and adding resistance to multiple drugs used for treatment.

In our reality, treatment is free, and the strategy of directly observed supervised treatment is used to reduce abandonment and improve treatment outcomes [2], with treatment for sensitive tuberculosis and a standardized treatment that is individualized

Á. Rocha et al. (Eds.): WorldCIST 2024, LNNS 986, pp. 199–210, 2024.
https://doi.org/10.1007/978-3-031-60218-4_19

based on sensitivity tests. In Peru, treatment for sensitive tuberculosis is divided into two phases, with the first consisting of four drugs taken daily except on Sundays for two months, completing 50 doses, and the second consisting of two drugs taken three times a week, isoniazid and rifampicin, for four months, completing 54 doses [3]. Thus, explaining the difficulty of concluding tuberculosis treatment.

Tuberculosis is a prominent global cause of mortality attributed to a singular infectious agent, resulting in over one million annual fatalities [4]. According to a reliable source, tuberculosis has the position of being the primary cause of death among infectious diseases on a global scale. Furthermore, the emergence of drug- resistant strains of tuberculosis presents a substantial risk to the overall security of global health. Although there has been a decline in their occurrence, Brazil and Peru continue to have one of the highest rates [6] in South America.

[7] conclude that during the years 2011 to 2015, progress was made in the control of tuberculosis, although it is still far from achieving the goal that by 2035, Peru will be free of this disease or have an incidence of less than 10 cases per 100,000 inhabitants [8].

One of the main obstacles is the rate of abandonment of treatment, which is influenced by the length of treatment, adverse reactions, living conditions, poverty, unemployment, and older age (And in their conclusion, [9] emphasize the partnership that should exist between tuberculosis control programs and primary care teams to reduce the risk factors associated with treatment abandonment and, as a result, multidrug resistance. Even with an organized and functional primary care, continuous education in tuberculosis and team participation in strategies for the control of this disease are required [10].

Strengthening adherence to treatment has a greater impact than other strategies, especially in developing nations, while others are geared toward the patient's needs and the identification of barriers to accessing treatment, with primary care being the most appropriate level of care for this management [11]. The spread of tuberculosis is often linked to health systems that lack stability, populations that are densely concentrated, and environments with inadequate ventilation. However, the abandonment of tuberculosis treatment is closely connected to the development of drug resistance. Consequently, it is crucial to identify patients who are at a higher risk of discontinuing treatment in order to mitigate the mortality rate associated with this disease [12]. The issue of treatment desertion in tuberculosis is a matter of public health concern, as it has significant implications for the well-being of patients, their families, and the wider community [13].

Tuberculosis continues to be the primary infectious cause of mortality on a global scale. Despite efforts to reduce its prevalence, the decline in incidence is not as rapid as expected. Consequently, the issue of drug resistance necessitates ongoing investigation and the development of effective solutions. Notably, treatment abandonment emerges as a significant contributor to this problem, leading to both drug resistance and heightened mortality rates.

A systematic review compared directly observed treatment by mouth with videoobserved therapy, reminders and markers, incentives and enablers, patient education, and staff education [15], highlighting the importance of preventing treatment abandonment. With the evaluation of indicators established by the tuberculosis control program, the possibility of underdiagnosis and community transmission is evident, and the incidence continues to fluctuate [16]. The short treatment strategy of direct observation has losses during its process for a variety of reasons, including access to medical care, distance and cost of transportation to the health center, moving within the past year, and absence of disease knowledge [17]. In our current reality, it is essential to implement strategies that identify the causes and preventative measures for patient abandonment.

The objective of this study is to identify the characteristics that are correlated with treatment abandonment among Tuberculosis patients in the Tacna Health Network.

2 Method

The study design employed in this research is an observational, retrospective, case-control, cross-sectional approach. The formula utilized in case-control research can be expressed as follows.

Case-control study

Risk or protective factor	Cases	Controls	
Exposed	a	b	a+b
Not exposed	c	d	c+d
	a+c	b+d	a+b+c+d

Proportion of exposed cases = a / (a+c)
Proportion of exposed controls = b / (b+d)

$FA = (OR - 1) / OR*$
$FAP = FA \times$ fraction of exposure in cases

*Formula valid for OR values similar to RR (rare diseases). Otherwise, $FA = (RR - 1) / RR$
RR can be estimated according to the following formula:
$RR = OR // ((1 - Prev) + (Prev \times OR))$

During 2018, the Institutional Development Office of the Tacna Health Network compiled information on 395 patients with a tuberculosis diagnosis. The sample consisted of 13 patients with a tuberculosis diagnosis who abandoned treatment during the study period and 117 controls who adhered to the tuberculosis treatment protocol.

Non-random sampling was performed out of expediency. The sample size is considered to be 130 cases who abandoned tuberculosis treatment in 2018 based on a ratio of 1 to 9 with the controls.

3 Results

(See Table 1).

Table 1. Determinants related to treatment abandonment in patients diagnosed with tuberculosis within the Tacna Health Network, with a specific focus on age.

Condition									
Tobacco use	Total Population		Case		Control		Association Test		
	$N° = 130\%$	%	$N° = 13$	$\%N° = 117\%$	$N° = 117$	%	X2P (Prob.)OR	P(Prob.)	OR
12 to 19 years	21	21,2%	1	0,8%	20	15,4%			
20 to 29 years	41	25,0%	5	3,8%	36	27,7%			
30 to 39 years	23	21,2%	1	0,8%	22	16,9%	2,867	2,867	0
40 to 49 years	24	24	4	3,1%	20	15,4%			
Older or equal to 50 years	1	15,4%	2	1,5%	19	14,6%			

The statistical analysis indicates that the ages were not statistically significant (p > 0.05); thus, age is not associated with treatment abandonment among tuberculosis patients (Table 2).

Table 2. Determinants related to treatment abandonment in patients diagnosed with tuberculosis within the Tacna Health Network, with a specific focus on sex

Condition								
Sex	Total Population		Case		Control	Association Test		
	$N° = 130$	%	$N° = 13$	%	$N° = 117\%$	X2	P (Prob.)	OR
Male	77	59,2%	8	6,2%	6953,1%	0,032	0,858	1,113
Female	53	40,8%	5	3,8%	4836,9%			

53.1% of patients who completed treatment were male, compared to 6.2% of patients who did not complete treatment; 36.9% of patients who completed treatment were female, compared to 3.8% of patients who did not complete treatment. The statistical analysis indicates that sex was not statistically significant (p > 0.05); therefore, sex is not associated with treatment abandonment among tuberculosis patients (Table 3).

Table 3. Determinants related to treatment abandonment in patients diagnosed with tuberculosis within the Tacna Health Network, with a specific focus on level of education

Condition							
Level of Education	Total Population		Case		Control		Association Test
	N° = 130	%	**N° = 13**	%	**N° = 117**	%	**X2P (Prob.)OR**
Illiterate	14	10,8%	2	1,5%	12	9,2%	
Primary	41	31,5%	7	5,4%	34	26,2%	4,7870,1880
Secondary	64	49,2%	4	3,1%	60	46,2%	
Highschool / Technical	11	8,5%	0	0,0%	11	8,5%	

It is shown with a higher frequency of 46.2% of patients who completed treatment were secondary and 3.1% of patients who did not complete treatment were secondary, followed by 26.2% of patients who completed treatment were primary and 5.4% of patients who did not complete treatment were primary and in the most cases a low percentage is observed. The statistical analysis shows that the degree of education was not significant (p > 0.05); therefore, the degree of education is not associated with treatment abandonment in patients with tuberculosis (Table 4).

Table 4. Determinants related to treatment abandonment in patients diagnosed with tuberculosis within the Tacna Health Network, with a specific focus on BMI.

Condition									
BMI	Total Population		Case		Control		Association Test		
	N° = 130	%	N° = 13	%	**N° = 117**	%	**X2P (Prob.) OR**		
Low weight (<18.5)	14	10,8%	1	,8%	13	10,0%			
Normal (18.5–24.9)	85	65,4%	10	7,7%	75	57,7%	1,293	0,731	0
Overweight	23	17,7%	1	,8%	22	16,9%			

(*continued*)

Table 4. (*continued*)

Condition									
BMI	Total Pop ulation		Case		Control		Association Test		
	N° = 130	%	N° = 13	%	N° = 117	%	X2P (Prob.) OR		
Low weight (<18.5)	14	10,8%	1	,8%	13	10,0%			
(25 – 29.9)									
Obese (≥ 30)	8	6,2%	1	,8%	7	5,4%			

Table 5. Determinants related to treatment abandonment in patients diagnosed with tuberculosis within the Tacna Health Network, with a specific focus on tobacco use.

Condition									
Tobacco use	Total Population		Case		Control		Association Test		
	N° = 130%		N° = 13	%N° = 117%	N° = 117	%	X2P (Prob.)OR	P(Prob.)	OR
Current	9	6,9%	2	1,5%	7	5,4%			
Past	25	19,2%	3	2,3%	22	16,9%	1,901	0,387	0
None	96	73,8%	8	6,2%	88	67,7%			

It is observed that 57.7% of patients who completed treatment had a normal BMI (18.5 - 24.9) and 7.7% of patients who did not complete treatment had a normal BMI (18.5 - 24.9); 16.9% of patients who completed treatment had an overweight BMI (25 - 29.9) and 0.8% of patients who did not complete treatment were overweight (25 - 29.9); and in the remaining cases, a low percentage is observed. The statistical analysis reveals that BMI was not statistically significant (p > 0.05); thus, BMI is not associated with treatment abandonment among tuberculosis patients (Table 5).

It is observed that 67.7% of patients who completed treatment in tobacco use were none and 6.2% of patients who abandoned treatment in tobacco use were none with a higher frequency, followed by 16.9% of patients who completed treatment in tobacco use being in the past and 2.3% of patients who abandoned treatment in tobacco use being in the past with a low frequency. Tobacco use is not associated with treatment abandonment among tuberculosis patients (Table 6).

With a higher frequency, 50.0% of patients who completed treatment in alcohol consumption did not consume any and 5.4% of patients who abandoned treatment in

Table 6. Determinants related to treatment abandonment in patients diagnosed with tuberculosis within the Tacna Health Network, with a specific focus on alcohol consumption.

Condition									
Alcohol consumption	Total Population		Case		Control		Association Test		
	N° = 130	%	N° = 13	%	N° = 117	%	X2	P (Prob.)	OR
Current	26	20,0%	2	1,5%	24	18,5%			
Past	32	24,6%	4	3,1%	28	21,5%	0,382	0,826	0
None	72	55,4%	7	5,4%	65	50,0%			

alcohol consumption did not consume any, followed by 21.5% of patients who completed treatment in alcohol consumption who consumed in the past and 3.1% of patients who abandoned treatment in alcohol consumption who were in the past. Alcohol consumption is not associated with treatment abandonment among tuberculosis patients according to statistical analysis (p > 0.05) (Table 7).

Table 7. Determinants related to treatment abandonment in patients diagnosed with tuberculosis within the Tacna Health Network, with a specific focus on drug use.

Condition									
Drugs use	Total Population		Case		Control		Association Test		
	N° = 130%		N° = 13	%N° = 117%	N° = 117	%	X2P (Prob.)OR	P(Prob.)	OR
Current	9	6,9%	1	0,8%	8	6,2%			
Past	7	5,4%	1	0,8%	6	4,6%	0,171	0,918	0
None	114	87,7%	11	8,5%	103	79,2%			

It is observed with the highest frequency that 79.2% of patients who completed treatment in drug use were none and 8.5% of patients who dropped out of treatment in drug use were none, followed by 6.2% of patients who completed treatment in drug use consuming currently and 0.8% of patients who dropped out of treatment in drug use consuming currently. The statistical analysis indicates that drug use was not statistically significant (p > 0.05); thus, drug use is not associated with treatment abandonment among tuberculosis patients.

4 Discussion

Although in our study, 27.7% of patients who completed treatment were between the ages of 20 and 29, and 3.8% of patients in the same age range abandoned treatment, we found no correlation between age and treatment abandonment. As was demonstrated in other studies [6, 18], there was no correlation between the patient's gender and treatment abandonment. In terms of marital status, 47.7% were unmarried, followed by cohabiting individuals, which had no bearing on treatment abandonment. In the educational level variable, 46.2% were high school graduates compared to 3.1% who did not complete treatment, without finding an association for treatment abandonment; in terms of employment status, the highest frequency was 36.2% of those who were unemployed but completed treatment, without finding an association for treatment abandonment; those who did not have children presented a frequency of 50.0% who completed treatment and 4.6% who did not complete treatment; and those who did not have children presented a frequency of 50.0% who completed treatment and 4.6% who did not complete treatment, without finding an association between this variable and treatment abandonment, showing that in these variables, unlike other studies, no association was found with treatment abandonment. [19]. According to the findings of a separate study [20], individuals who possess a lower level of education, belong to the male gender, identify as black, are institutionalized patients, or suffer from pulmonary and extrapulmonary tuberculosis exhibit a higher propensity for treatment non-adherence. In a research study aimed at identifying individuals who are at risk of dropping out based on their individual health determinants, the findings revealed certain demographic characteristics. Specifically, it was observed that a significant proportion of individuals at risk of dropping out were men (20.0%), had black skin (20.3%), fell within the age range of 20 to 39 years (21.8%), had completed between 4 and 7 years of schooling (23.6%), had reentered treatment after previously dropping out (36.5%), engaged in alcohol consumption (31.0%), used drugs (39.3%), were smokers (26.5%), and experienced homelessness (55.4%). Additionally, the study examined the ecological characteristics and found that individuals residing in municipalities with high human development index and income inequality exhibited a higher likelihood of dropout [21]. If we examine the clinical epidemiological factors, the normal body mass index is found in 55.7% of those who completed treatment and 7.7% of those who did not complete treatment, demonstrating no correlation with treatment abandonment. As for tobacco consumption, 67.7% of those who completed treatment did not use tobacco, while 6.2% abandoned treatment. If we relate alcohol consumption to treatment completion, 50.0% of those who completed treatment had no consumption and 5.4% dropped out of treatment, but no association was found; 79.2% of those who completed treatment had no drug consumption and 9.5% dropped out of treatment, but unlike other studies, no association was found for dropout [22]. In a risk stratification study, factors for treatment discontinuation, cigarette smoking, substance use, continued admissions, and a high dropout risk score were identified, but only for smoking cessation [23]. In another study [24], active smokers had a lower chance of cure (62.1% versus 82.5%; p = 0.032) and a higher chance of ceasing (31.0% versus 12.7%; p = 0.035).

When it comes to overcrowding, 73.8% of those who completed treatment did not have overcrowding, and 8.5% dropped out, with no correlation between overcrowding and dropout; 54.6% of patients with a family history of tuberculosis completed treatment, and 5.4% dropped out, with no correlation between the history and dropout; and 59.2% of those who completed treatment had a positive initial smear test. Regarding the duration of illness from 1 to 8 weeks, 84.6% of those who completed treatment and 8.5% of those who abandoned treatment were found not to be statistically significant for treatment abandonment, demonstrating that there was no association with clinical epidemiological factors in our study.

In aspects of knowledge there was evidence of lack of it about the disease and absence of knowledge about the treatment [25], showed different results to our study.

In a different study of unfavorable outcomes of tuberculosis treatment in adolescents and young adults, non-similar results were discovered highlighting the lack of housing, HIV and illicit drug use, and government support reduces the incidence of unfavorable outcomes [26].

If we review the barriers to accessing medical care, we find that in our study, 33.8% of those who completed treatment and 0.8% of those who abandoned treatment received an evaluation lasting longer than one week, indicating a correlation between the evaluation and treatment abandonment.

In medical evaluation, nurse, nutritionist, psychologist, and social worker, 51.5% of patients who completed treatment had an evaluation after one week, while 1.5% abandoned treatment; there was a correlation between this variable and abandonment. Similar to other studies, it is essential to analyze treatment irregularity, as 29.2% of those who completed treatment did not exhibit any and 0.8% of those who abandoned treatment did not. In phase one, an association was found between treatment abandonment and treatment irregularity.

In phase two, the most common duration of non-attendance was between 1 and 20 days, 48.5% completed treatment, and 3.1% abandoned treatment, indicating a correlation between non-attendance and treatment abandonment. Most frequently, between the absence and the first visit, there were no visits. This was the case for 52.2% of those who completed treatment and 2.5% of those who fell out, indicating a correlation between treatment completion and treatment abandonment.

In addition, 46.2% of those who completed treatment had one visit during phase one, whereas only 5.4% of those who abandoned treatment did, indicating a significant correlation. During phase two, 56.2% of those who completed treatment did not attend any follow-up appointments, whereas 5.4% of those who fell out were associated with treatment abandonment.

Distinct studies have found a correlation with barriers to accessing medical care [27], whereas another study found no correlation [18] as a result of distinct methodologies.

In our study, care by the multidisciplinary team, patient follow-up through home visits, and non-attendance at treatment, as well as early identification of treatment irregularities, were predominant. In our health care system, it is crucial to continue the good actions we have been taking, but it is also essential to evaluate whether they are being carried out with the same level of excellence as before.

5 Conclusions

When conducting a comparative analysis between the reasons for prescription non-adherence among patients in the Tacna Health Network and previous research that primarily examined alcohol, tobacco, and other substance use, noticeable changes in patterns emerge.

In the Tacna Health Network, no statistically significant correlation was seen between age, sex, marital status, education level, employment position, or family size and the choice to discontinue tuberculosis treatment. A study conducted on patients within the Tacna Health Network found no significant association between the discontinuation of tuberculosis therapy and many factors, including body mass index, overcrowding, family history of tuberculosis, baseline smear microscopy, and duration of illness.

This study identified several factors associated with the abandonment of tuberculosis treatment among patients of the Tacna Health Network. These factors included the inclusion of a medical evaluation nurse nutritionist, psychologist, and social worker in the treatment process, as well as irregularities in the treatment phases. Additionally, non-attendance during both phase one and phase two, the duration between non-attendance and the initial visit, and the frequency of visits during both phases were also found to be associated with treatment abandonment.

References

1. Bonilla, C.: Factores de riesgo asociados al abandono del tratamiento en pacientes con tuberculosis multidrogorresistente en la región callao, Perú, años 2010–2012. [Tesis de maestría, Universidad Peruana Unión]. Repositorio de la Universidad Peruana Unión (2016). https://repositorio.upeu.edu.pe/handle/20.500.12840/656
2. Zevallos, M.: Factores asociados al abandono del tratamiento antituberculoso esquema i en la red de salud San Juan de Lurigancho, Lima, Perú [Tesis de maestría]. Universidad Peruana Cayetano Heredia (2017). https://reposito-rio.upch.edu.pe/bitstream/handle/20.500.12866/1030/Factores_ZevallosRomero_Maritza.pdf?sequence=1&isAllowed=y
3. Alarcón, V., et al.: Norma técnica de salud para la atención integral de las personas afectadas por tuberculosis, vol. 1, Ministerio de salud (2013). http://www.tuberculosis.minsa.gob.pe/portaldpctb/recursos/20180308083418.pdf
4. Ledesma, J., Ma, J., Vongpradith, A., Maddison, E., Novotney, A., Biehl, M., et al.: Global, regional, and national sex differences in the global burden of tuberculosis by HIV status, 1990–2019: results from the global burden of disease Study 2019. Lancet Infect. Dis. **22**(2), 222–241 (2021). https://doi.org/10.1016/S1473-3099(21)00449-7
5. Floyd, K., Glaziou, P., Zumla, A., Raviglione, M.: The global tuberculosis epidemic and progress in care, prevention, and research: an overview in year 3 of the End TB era. Lancet Respir Med, **6**(4), 299–314 (2018). https://doi.org/10.1016/S2213-2600(18)30057-2, https://www.scielo.br/j/csp/a/bnNQf4rdcMNpPjgfnpWPQzr/?lang=en&format=html
6. Anduaga, A., et al.: Factores de riesgo para el abandono del tratamiento de tuberculosis pulmonar sensible en un establecimiento de salud de atención primaria, Lima, Perú. Acta Médica Peru, **33**(1), 21–8 (2016). http://repebis.upch.edu.pe/articulos/acta.med.per/v33n1/a5.pdf
7. Alarcón, V., Alarcón, E., Figueroa, C., Mendoza-Ticona, A.: Tuberculosis en el Perú: situación epidemiológica, avances y desafíos para su control. Rev. Peru Med. Exp. Salud Publica **34**(2), 299–310 (2017). https://doi.org/10.17843/rpmesp.2017.342.2384

8. Cáceres, F.: Factores de riesgo para abandono (no adherencia) del tratamiento antituberculoso. MedUNAB, **7**(21), 172–80 (2004). https://revistas.unab.edu.co/index.php/medunab/article/view/215

9. Fregona, G., et al.: Risk factors associated with multidrug-resistant tuberculosis in Espírito Santo. Brazil. Rev Saude Publica **51**(41), 1–11 (2017). https://doi.org/10.1590/S15188787.2017051006688

10. Costa, J., et al.: Performance assessment of primary healthcare services in tuberculosis control in a city in Southeast Brazil. Cad Saude Publica **37**(3), 1–13 (2021). https://doi.org/10.1590/0102311x00112020

11. Navarro, P., et al.: The impact of the stratification by degree of clinical severity and abandonment risk of tuberculosis treatment. J. Bras. Pneumol. 47(4), 1–9 (2021). https://doi.org/10.36416/18063756/e20210018

12. Da Frota, V., Bastos, A., Vieira, I., Gimeniz, M.: Aspects associated with drug resistance in people with tuberculosis/HIV: An integrative review. ACTA Paul Enferm, **33**(0), 1–8 (2020). https://doi.org/10.37689/acta-ape/2020AR01316

13. Rivera, O., Santiago, J.: Abandono del tratamiento en tuberculosis multirresistente: factores asociados en una región con alta carga de la enfermedad en Perú. Biomedica **39**(2), 44–57 (2019). https://doi.org/10.7705/biomedica.v39i3.4564

14. Carneiro, G., De Oliveira, A., De Holanda, E., De Vasconcelos, E., Dos Santos, C., Ramos, V.: Priority areas for the control of tuberculosis treatment abandonment in Recife. Brazil. Mundo da Saude **45**(1), 210–220 (2021). https://doi.org/10.15343/0104-7809.202145210220

15. Alipanah, N., et al.: Adherence interventions and outcomes of tuberculosis treatment: a systematic review and meta-analysis of trials and observational studies. PLoS Med. **15**(7), 1–44 (2018). https://doi.org/10.1371/journal.pmed.1002595

16. Jam, M., León, Y., Sierra, D., Jam, B.: Tuberculosis pulmonar: estudio clínicoepidemiológico. Revista Cubana de Medicina General Integral **33**(3), 321–330 (2017). https://www.mendeley.com/catalogue/5931e359-cf6a-3833-bfa0-b6d877862776/?ref=raven&dgcid=raven_md_suggest_email&dgcid=raven_md_suggest_mie_email

17. Ruru, Y., et al.: Factors associated with non-adherence during tuberculosis treatment among patients treated with DOTS strategy in Jayapura, Papua Province, Indonesia. Glob. Health Action **11**(1), 1–8 (2018). https://doi.org/10.1080/16549716.2018.1510592

18. Sinchi, G.: Factores asociados al abandono de tratamiento antituberculosis esquema I en el centro de salud defensores de la patria, ventanilla 2017–2019 [Tesis de grado, Universidad Privada San Juan Bautista]. Repositorio de la Universidad Privada San Juan Bautista (2020). http://repositorio.upsjb.edu.pe/handle/upsjb/2865

19. Llerena, Y.: Factores que inducen al abandono de tratamiento en pacientes con tuberculosis del centro de Salud Ampliación Paucarpata - Arequipa 2015 [Tesis de grado, Universidad Alas Peruanas]. Repositorio de la Universidad Alas Peruanas (2017). https://hdl.handle.net/20.500.12990/357

20. Medeiros, M., Castro, N., Paes, A.Y., Pires, L.: Aspectos sociodemográficos e clínicoepidemiológicos do abandono do tratamento de tuberculose em Pernambuco, Brasil, 20012014. *Epidemiol.* eServ. Saude **26**(2), 369–78 (2017). https://doi.org/10.5123/S16794974201700 0200014

21. Lima, S.V.M.A., de Araújo, K.C.G.M., Nunes, M.A.P. and Nunes, C.: Early identification of individuals at risk for loss to follow-up of tuberculosis treatment: A generalised hierarchical analysis. *Heliyon,* **7**(4) (2021). https://doi.org/10.1016/j.heliyon.2021.e06788

22. Salvador, M.: Frecuencia y factores asociados al abandono del tratamiento en pacientes con tuberculosis pulmonar en la provincia de Ica, Perú, 2015–2019 [Tesis de grado, Universidad César Vallejo]. Repositorio de la Universidad César Vallejo (2020). https://hdl.handle.net/20.500.12692/56101

23. Peres, W., et al.: Risk stratification and factors associated with abandonment of tuberculosis treatment in a secondary referral unit. Patient Prefer. Adherence **14**(1), 2389–2397 (2020). https://doi.org/10.2147/ppa.s266475

24. De Vargas, K., Freitas, A., Azeredo, A.Y., Silva, D.: Smoking prevalence and effects on treatment outcomes in patients with tuberculosis. Rev. Assoc. Med. Bras. **67**(3), 406–410 (2021). https://doi.org/10.1590/1806-9282.20200825

25. Mansour, G.K., Ferreira, L.D.P.Q., de Oliveira Martins, G., Melo, J.L.L., Freitas, P.S., do Nascimento, M.C.: Factors Related to Non-Adherence and Abandonment of Pulmonary Tuberculosis Treatment Type of Study. Medicina Ribeirão Preto **54**(2) 1–12 (2021). https://doi.org/10.11606/issn.2176-7262.rmrp.2021.172543

26. Chenciner, L., Annerstedt, K., Pescarini, J., Wingfield, T.: Social and health factors associated with unfavourable treatment outcome in adolescents and young adults with tuberculosis in Brazil: a national retrospective cohort study. Lancet Glob Heal **9**(10), 1380–1390 (2021). https://doi.org/10.1016/S2214-109X(21)00300-4

27. Llanos, J., Trujillo, R.: Factores asociados al abandono de tratamiento en personas afectadas por tuberculosis en la micro red, Chiclayo 2015 [Tesis de Licenciatura, Universidad Señor de Sipán]. Repositorio de la Universidad Señor de Sipán (2015). https://hdl.handle.net/20.500.12802/532

Comprehensive Analysis of Feature Extraction Methods for Emotion Recognition on Motor Imagery from Multichannel EEG Recordings

Amr F. Mohamed[(✉)] [iD] and Vacius Jusas [iD]

Kaunas University of Technology, Studentų 50, 51390 Kaunas, Lithuania
amr.mohamed@ktu.edu, vacius.jusas@ktu.lt

Abstract. Electroencephalogram (EEG) related analyses have a vast extent of enquiries, preliminary from Brain-Computer Interfaces (BCI) on motor imagery tasks, neurotechnology, human-computer interaction, clinical and medical applications, sleep research, as well as the neural basis of emotional recognition and behavioural responses. This study purposely investigates the application of feature extraction methods from the emotion recognition arena against the motor imagery arena, using the electroencephalogram (EEG) data from the BCI Competition IV. As the same methods are used in both fields, binary classification has been embraced, emotion recognition's valence and arousal, using its equivalent motor imageries of left and right classes. Essentially, six EEG feature extraction methods were used in the classification accuracy process, including statistical features, wavelet analysis, higher-order spectra (HOS), Hjorth, fractal dimension (Katz, Higuchi, Petrosian) and three-dimensional combination of fractal, wavelet and Hjorth. Further, the classifier performance methods were the Gaussian radial basis function RBF SVM (GSVM) and the regression tree (CART). Remarkably, the statistical method has better accuracy than the other feature sets, precisely the fractal dimension, which was the highest considering the emotion recognition span. Also, GSVM generally has better accuracy on the BCI IV dataset than CART. By contemplating the novel outcome notices, we can deduce that in the methods used in emotion recognition when applied to motor imagery, unique results are obtained, feature-wise and classifier-wise.

Keywords: Electroencephalogram · Emotion Recognition · Motor Imagery · EEG Feature Extraction · Brain-Computer Interface

1 Introduction

When measuring and monitoring brain activity, various methods can be used, like functional magnetic resonance imaging (fMRI), functional near-infrared spectroscopy (fNIRS), and electroencephalogram (EEG). The latter is the most common technique globally used by scientists in related fields of study, as it has no intrusiveness, no surgical operations are required, is less expensive when compared to the other techniques, and all is done using a brain-computer interface scalp placed on the head of the subject.

EEG is the approach used to record electrical signals from the brain as the brain emits electrical charges. EEG data is being acquired using sensors and electrodes in the BCI device. There are numerous use cases of EEG and how it can be used, detecting abnormal brain activity in clinical diagnosis, which can help with epilepsy or seizures. The study and investigations of emotional recognition procedures, with different emotional states (i.e., valence, arousal) and how they influence the behavioural interaction of the subject with his surroundings. Brain-computer interfaces, on the other hand, allow a physically disabled person to perform motor imagery tasks and control an external electronic device using their brain, such as imagining the movement of the right and left hand, both feet and tongue.

The key focus will be emotion recognition (ER-EEG) and motor imagery task (MI-EEG) use cases. These two fields have seen a lot of research year to date. Yet, there is still room for innovation in both fields.

This study will examine an original notion by combining EEG-related emotion recognition and motor imagery studies. As both fields of research have a vast area of feature extraction and performance classification techniques, the novel idea is to apply emotion recognition feature extraction techniques on motor imagery datasets, as the same methods are being used in both fields. The fundamental contribution of this study is to assess how the EEG data is being interpreted in both spans; if the same feature sets and classifiers are being used, it will have the similar or dissimilar consequences.

2   Related Work

The present study will use a mixed ER-EEG and MI-EEG pipeline exploiting the BCI Competition IV 2a motor imagery dataset.

Commencing with the emotion recognition pipeline, the latter has been implemented in a recent research article [1] that was applied to Emotion-Related EEG datasets: MAHNOB-HCI, DEAP (Dataset for Emotion Analysis using Physiological signals), SEED (SJUT emotion EEG Dataset), AMIGOS (A dataset for Mood, personality, and affect research on individuals and Groups), and DREAMER. Firstly, the MAHNOB-HCI was established in 2011 by Soleymani et al. [2]; a total of 32-channel EEG recordings were obtained from 27 participants while viewing 20 video clips that lasted in a range of 34,9 s to 117 s, with a sampling frequency of 256 Hz. Secondly, the DEAP dataset was established in 2012 by [3]; similarly, 32-channel EEG recordings were obtained from 32 participants while viewing 40 music video clips, each lasting a minute in total, with a 512 Hz frequency. Thirdly, in the SEED dataset, established in 2015 by Zheng et al. [4], 62-channel EEG recordings were obtained from 15 participants while viewing 10 videos that lasted approximately 240 s in three sessions, with a sampling rate of 1000 Hz. Fourthly, the AMIGOS dataset was established by [5] in 2018, with 40 participants; 14-channel EEG recordings were obtained from 40 participants viewing 16 film clips, which lasted 250 s, in 2 sessions and had a sampling frequency of 128 Hz. Lastly, the DREAMER dataset was developed by [6] in 2018; 14-channel EEG recordings were obtained from 23 participants, viewing 18 video clips with a length ranging from 65 to 393 s, one session in total, and a sample frequency of 128 Hz. During the study of the mentioned article, a two-dimensional model of emotions, the valence-arousal plane, was used. Not all datasets have a binary classification; the emotional states selected were valence and arousal or positive and negative. These ER-EEG datasets were transformed into a self-report scale (ground truth). They underwent a 4-dimensional preprocessing phase: notch filter, high-pass filter, down-sampling, and common average reference (CAR) montage. This is followed by the feature extraction phase, including statistical, wavelet, fractal dimension, Hjorth parameters, higher order spectra and a combination of all the previously mentioned feature sets [7–11]. Then, two principal performance measures, the Gaussian radial basis function RBF SVM (GSVM) and the regression tree (CART) classify emotional states, valence, and arousal as high or low. During that study, a deduction was made stating that the CART classifier performed better for EEG emotion recognition compared to GSVM and the feature set, fractal dimension, namely Katz [12], Petrosian [13], and Higuchi [14], achieved the best overall mean accuracy for binary classification of valence and arousal.

Proceeding with the motor imagery pipeline, implemented in the article [15] that was applied to the Motor Imagery-Related EEG dataset: the BCI Competition IV 2a and 2b dataset [16]. The 2a dataset has a total of 25 electrodes that were used; the first 22 are EEG, and the last 3 are EOG (Electrooculography). The sampling frequency is 250 Hz. It was applied on 9 subjects with 9 runs (from 1 to 3 specified for eye movement and from 4 to 9 specified for motor imagery). The related event types, the 4 MI classes are cue onset left, right, foot and tongue. The mentioned article respects a common filter bank spatial pattern that applies band-pass and spatial filtering as a preprocessing phase. Regarding feature selection, applying CSP (covariance and composite covariance, white matrix and common eigenvector for binary classes left and right), mutual information-based best individual feature and mutual information-based rough set theory. The support vector machine (SVM) model was used as a classifier and yielded an original 10×10-fold cross-validation classification accuracy of 89.5% averaged over the 5 subjects, however the latest FBCSP algorithm yielded a 10×10-fold cross-validation classification accuracy of 90.3% [15].

The current work being established in this paper is noticeably distinctive compared to the previously mentioned existing research, whether related to emotion recognition or motor imagery tasks. Applying ER-EEG-related feature sets and classifiers to MI-EEG can differentiate how scientists interpret EEG signals when it comes from emotions like valence or arousal or when it comes from motor imagery tasks like imagining the movement of an arm or a leg. By examining if the same methods will affect the overall results similarly or differently, during their application to both spans.

3 Background and Theory

This section presents any foundational concepts, definitions, or relevant theories that will be used in the current study and must be comprehended, i.e., preprocessing filters, feature extraction methods, performance classifiers, etc.

3.1 Notch Filtering

Notch filtering [21] typically removes a specific frequency from an EEG signal. It is designed to achieve a narrow, deep cut called a notch at a specified frequency. Using the latter will eliminate any noise from the signals in the study whilst considering its parameters: the sampling frequency Fs, the powerline frequency $f0$, and the quality factor Q. The mathematical representation of a digital notch filter can be represented by a transfer function $H(z)$, given by (1):

$$H(z) = \frac{Y(z)}{X(z)} = \frac{\sum_{m=0}^{M} b_m z^{-m}}{1 + \sum_{n=1}^{N} a_n z^{-n}} \tag{1}$$

where:

- b_m are the feedforward coefficients (numerator),
- a_n are the feedback coefficients (denominator),
- M is the order of the numerator,
- N is the order of the denominator,
- $X(z)$ is the Z-transform of the input signal,
- $Y(z)$ is the Z-transform of the output signal.

3.2   High-Pass Filtering

High-pass Butterworth filtering (HPF) [22] allows high-frequency EEG signals to pass through while reducing amplitude. It is commonly used to remove low-frequency noise without affecting the higher-frequency brainwaves that are meant to remain. It is useful for reducing artefacts, dealing with electrode drifts when recording EEG signals from a BCI (scalp) device, and enhancing the overall quality of the EEG signals in study. The mathematical formula for a Butterworth high-pass filter with parameters of order N and with a cutoff frequency ωc is generally expressed as follows (2):

$$H(z) = \frac{B(z)}{A(z)} \tag{2}$$

where $B(z)$ and $A(z)$ are the Z-transforms of the filter coefficients b and a, respectively. In our case we will be using a fourth order HPF filter, with a transfer function that will use b_i and a_i, representing the filter coefficients, given by (3):

$$H(z) = \frac{b_0 + b_1 z^{-1} + b_2 z^{-2} + b_3 z^{-3} + b_4 z^{-4}}{1 + a_1 z^{-1} + a_2 z^{-2} + a_3 z^{-3} + a_4 z^{-4}} \tag{3}$$

Consequently, we calculate the coefficients b and a for the fourth-order high-pass Butterworth filter with the normalised cutoff frequency ωc. To apply the previously mentioned filter to our EEG signals, we use multiplication in the Z-domain, given by (4):

$$Filtered\ Signal(z) = H(z) \cdot EEG\ Signal(z) \tag{4}$$

3.3 Down-Sampling Filter

Downsampling [23] is a process used to reduce the sampling rate of EEG signals. The key purpose is to reduce noise in the signals and efficiently reduce the data size, which decreases the computational load for the processing and analysis of the data. We can express the downsampling process mathematically through the following steps.

- Downsampling factor calculation is the ratio of the original sampling frequency $F_{original}$ to the target sampling frequency F_{target}, , mathematically given by (5):

$$D = \left\lfloor \frac{F_{original}}{F_{target}} \right\rfloor \tag{5}$$

- The EEG signal is resampled to a new length determined by the original length of the signal divided by the previously mentioned downsampling factor. Hence, if we assume L is the length of the original signal, then the new L' would be given by (6):

$$L' = \frac{L}{D} \tag{6}$$

- We can generally represent the overall downsampling operation in general form, given by (7):

$$EEG_{downsampled}(n) = RESAMPLE(EEG_{original}(t), L') \tag{7}$$

where: $EEG_{original}(t)$ is the original EEG signal, $EEG_{downsampled}(n)$ is the down sampled EEG signal, t represents the continuous time index, and n represents the discrete time index after downsampling.

3.4 CAR Montage Filter

The common average reference (CAR) montage [24] is a technique used in EEG signal processing that minimises the effect of common noise across all EEG channels. Enhances the brain's electrical activity detection relative to the average reference. The signal-to-noise ratio and the overall quality of the EEG data are improved whilst using the latter. Let's assume that the matrix X represents the EEG signal, where each row corresponds to a sample time, and each column corresponds to a channel. The CAR operation can be represented mathematically given by (8):

$$CAR = X - \frac{1}{N} \sum_{i=1}^{N} X_i \tag{8}$$

where: CAR is the EEG data matrix after the CAR montage has been applied, X is the original EEG data matrix, N is the number of EEG channels, and X_i is the signal of the i-th channel.

3.5   Statistical Features

Statistical features [7] are commonly used in EEG signal processing. Often used in brain-computer interfaces and seizure detection applications. It has four main measures; mean, variance, skewness, and kurtosis. Firstly, the mean ($\bar{x}$) value, which is the average value of the signal and is given by (9):

$$\bar{x} = \frac{1}{N} \sum_{i=1}^{N} x_i \tag{9}$$

where: N is the number of samples and x_i is the i-th sample in the signal. Secondly, the variance (σ^2) which measures the spread of the data points around the mean and is given by (10):

$$\sigma^2 = \frac{1}{N} \sum_{i=1}^{N} (x_i - \bar{x})^2 \tag{10}$$

Thirdly, the skewness which measures the asymmetry of the probability distribution of a real-valued random variable in comparison of its mean. It can be given by (11):

$$\text{Skewness} = \frac{1}{N} \sum_{i=1}^{N} \left(\frac{x_i - \bar{x}}{\sigma}\right)^3 \tag{11}$$

where: σ is the standard deviation of the signal. Lastly, the kurtosis measures if the tails of a given distribution contain any extreme values, like the skewness which considers the probability distribution of a real-valued random variable. It can be given by (12):

$$\text{Kurtosis} = \frac{1}{N} \sum_{i=1}^{N} \left(\frac{x_i - \bar{x}}{\sigma}\right)^4 - 3 \tag{12}$$

where: the subtraction by 3 is performed to make the kurtosis of the normal distribution zero.

3.6   Wavelet Analysis Features

The wavelet transforms [17] is used in EEG data analysis because it decomposes a signal into different frequency components while retaining time information. It has three main features: energy, entropy, and variance. Firstly, the energy, which represents the square sum of the wavelet coefficients and measures the signal's power at various scales, it can be given by (13):

$$\text{Energy} = \sum_{i=1}^{N} c_i^2 \tag{13}$$

where: c_i are the wavelet coefficients, and N is the total number of coefficients. Secondly, the entropy that measures the complexity of the EEG signal by quantifying the randomness in the distribution of the coefficients is given by (14):

$$\text{Entropy} = -\sum_{i=1}^{M} p_i \log(p_i) \tag{14}$$

where: p_i is the probability of the i-th coefficient, and M is the number of distinct coefficient values. Lastly, the variance, which measures the dispersion of the wavelet coefficients around their mean, given by (15):

$$\text{Variance} = \frac{1}{N-1} \sum_{i=1}^{N} (c_i - \bar{c})^2 \tag{15}$$

where: $\bar{c}$ is the mean of the wavelet coefficients.

3.7 Higher Order Spectra Features

The higher order spectra features [20] are part of the power spectral density (PSD), which estimates the power of an EEG signal's variation, it helps understand the distribution of electrical activity occurring at different frequencies. HOS features that will be considered during this study are skewness and kurtosis. Firstly, skewness, provides a measure of the symmetry of the power spectral density around its mean, and the kurtosis provides a measure for the specific distribution form of the power spectral density function. Both can be given by (16, 17):

$$\text{Skewness} = \frac{\frac{1}{N} \sum_{i=1}^{N} (Pxx_i - \overline{Pxx})^3}{\left(\sqrt{\frac{1}{N} \sum_{i=1}^{N} (Pxx_i - \overline{Pxx})^2}\right)^3} \tag{16}$$

$$\text{Kurtosis} = \frac{\frac{1}{N} \sum_{i=1}^{N} (Pxx_i - \overline{Pxx})^4}{\left(\frac{1}{N} \sum_{i=1}^{N} (Pxx_i - \overline{Pxx})^2\right)^2} - 3 \tag{17}$$

where: Pxx_i represents the power spectral density values, and the $\overline{Pxx}$ is the mean of those values, and N is the number of observations in the PSD.

3.8 Hjorth Features

Hjorth parameters [18, 19] are statistical measures used to characterize the time-domain EEG signals in terms of three principal aspects: activity, mobility, and complexity. These parameters can provide insights into the energy, frequency, and irregularity of the EEG signal. Firstly, activity parameter A is the variance of the EEG signal, given by the Eq. (18):

$$A = \text{Var}(x(t)) = \frac{1}{N} \sum_{i=1}^{N} (x_i - \bar{x})^2 \tag{18}$$

where: x_i are the samples of the signal, $\bar{x}$ is the mean of signal, and N is the number of samples. Secondly, mobility parameter M which is the square root of the variance of the first derivative of the EEG signal normalized by the activity, given by (19):

$$M = \sqrt{\frac{\text{Var}(\Delta x(t))}{A}} = \sqrt{\frac{\frac{1}{N-1} \sum_{i=1}^{N-1} (\Delta x_i)^2}{A}} \tag{19}$$

where: $\Delta x(t)$ is the first derivative of $x(t)$, and $\Delta x_i = x_{i+1} - x_i$ are the first differences of the signal. Lastly, complexity parameter C, which is the ratio of the mobility of the first derivative of the signal to the mobility of the EEG signal itself. Given by (20):

$$C = \frac{M_{\Delta x(t)}}{M} = \frac{\sqrt{\frac{Var(\Delta^2 x(t))}{Var(\Delta x(t))}}}{M} \tag{20}$$

where: $\Delta^2 x(t)$ is the second derivative of $x(t)$, and $\Delta^2 x_i = \Delta x_{i+1} - \Delta x_i$ are the second differences of the signal.

3.9 Fractal Dimension Features

Currently, we have considered several fractal dimension algorithms frequently used in EEG signal analysis, namely Katz, Petrosian and Higuchi. The fractal dimension provides insights into the self-similarity or complexity of the signal. Firstly, The Katz fractal dimension [12] is based on the idea of measuring how the signal deviates from being perfectly smooth. It can be given by (21):

$$FD_{Katz} = \frac{\log(n)}{\log(d) + \log(\frac{n}{a})} \tag{21}$$

where: log is the natural logarithm, n is the number of observations in the time series, d is the maximum change between two consecutive data points (the geometric complexity of the time series), and a is the distance from the first point to the last point (like the linear size of the time series). Secondly, the Petrosian fractal dimension [13] is based on the idea of measuring the irregularity or self-similarity of the signal by counting the number of sign changes in its first derivative. It can be given by (22):

$$FD_{Petrosian} = \frac{\log_{10}(n)}{\log_{10}(n) + \log_{10}\left(\frac{n}{n+0.4 \cdot N_\delta}\right)} \tag{22}$$

where: $\log_{10}$ is the base-10 logarithm, n is the total number of points in the time series, N_δ is the number of sign changes in the time series derivative, which gives an indication of the frequency content and waveform complexity. Lastly, the Higuchi fractal dimension [14] which measures the roughness or complexity of the signal by considering the changes in the signal as it is progressively downsampled. It can be given by (23, 24, 25):

$$HFD = -slope(\log(x), \log(L)) \tag{23}$$

where:

$$L_k = \frac{1}{k} \sum_{m=0}^{k-1} \left(\frac{N-1}{k(N-m-1)} \sum_{i=1}^{\frac{N-m}{k}} |data[m+ik] - data[m+(i-1)k]| \right) \tag{24}$$

and:

$$x = [1, 2, \ldots, k_{max}] \tag{25}$$

HFD, this is the outcome of the computation. L_k, this is an average measure over k sets. k is an integer that defines the time interval for the calculation of L_k. k_{max} is the maximum value of k that determines the number of different scales at which the time series is analyzed. N is the total number of data points. m is the index that runs from 0 to $k-1$. L_{mk}, represents the length of the curve for a particular scale k and a particular point m. i is the index used to sum over $N - m$ data points. *Data* refers to the actual time series EEG data points being analyzed. And x is the list of integers from 1 to k_{max} which is used along with L to compute the slope.

3.10 GSVM Classifier

The Gaussian Radial Basis Function (RBF) support vector machine (GSVM) [25] is a machine learning model that uses the Radial Basis Function kernel, also known as the Gaussian kernel, in a Support Vector Machine framework. It is useful when dealing with non-linearly separable data in EEG signal processing. The Gaussian RBF kernel is defined as follows, given by (26):

$$K(x, x') = \exp\left(-\gamma |x - x'|^2\right) \tag{26}$$

where: K is the RBF kernel function, $\mathbf{x}$ and $\mathbf{x}'$ are two feature vectors in the input space, γ is a parameter that defines how much influence a single training example has. The larger the γ is, the closer other examples must be to affect the classification. In the context of EEG, each $\mathbf{x}$ might represent a vector of EEG signal features (like frequency, band powers, wavelet coefficients, etc.), and $\mathbf{x}'$ would be another vector of features. The kernel function computes a similarity between these feature vectors that the SVM uses to classify EEG signals. The decision function of the SVM in this kernel-induced feature space can be represented as, given by (27):

$$f(x) = \sum_{i=1}^{N} \alpha_i y_i K(x, x_i) + b \tag{27}$$

where: N is the number of support vectors, α_i are Lagrange multipliers obtained by solving the SVM optimisation problem, y_i are the class labels of the training examples, and b is the bias term.

3.11 CART Classifier

The Classification and Regression Trees (CART) [26] algorithm is a popular decision tree-based machine learning method used for both classification and regression tasks. In the context of EEG (electroencephalogram) signal analysis, CART can be used to classify different mental states or detect patterns associated with specific neurological conditions. It can be described in four key steps. First, we begin with the entire set of EEG data as the root node. Then, we choose a variable and a split point that best separates the EEG data into two groups to create child nodes. The 'best' split is often determined by the split that maximizes the homogeneity of the target variable within each child node. And repeat the splitting process on each child node until a stopping criterion is met (e.g., maximum tree depth, minimum node size, or a minimal decrease in impurity). It can be given by the following pseudocode:

Given a set of EEG features X and a target variable Y:

1. Start with the root node containing all instances.
2. If all instances have the same value for Y, stop. Otherwise, proceed.
3. Select feature x_i and threshold θ to split on:

$$x_i, \theta = \arg\min_{x,t} \text{Impurity}(X, Y, x, t)$$

4. Split the node into two child nodes:

$$X_{\text{left}} = \{x \in X | x_i \leq \theta\}$$
$$X_{\text{right}} = \{x \in X | x_i > \theta\}$$

5. Repeat steps 2–4 recursively for X_{left} and X_{right}.
6. Stop if a maximum tree depth is reached or if further splitting does not improve impurity measures significantly.

where: X is the set of input variables (EEG features), Y is the target variable (e.g., type of brain activity), x_i is a feature from the EEG features, θ is the threshold value for splitting a node, Impurity(X, Y, x, t) is a measure of the homogeneity of the target variable within the nodes after the split. It can be Gini impurity, entropy, or another suitable metric, X_{left} is a subset of the data where the value of x_i is less than or equal to θ, X_{right} is a subset of the data where the value of x_i is greater than θ and $\arg\min_{x,t}$ is the argument of the minimum; the values of x and t that minimize the impurity.

4   Methodology

This section presents the description of the dataset being used, the architectural design, and the implementation of the proposed system.

4.1   Dataset

Throughout this study, our focus will be on the BCI Competition IV 2a dataset [16]. The latter was released as part of the fourth BCI Competition and contains EEG recordings from 9 subjects performing different motor tasks. It includes 25 channels, a sampling rate of 250 Hz and a bandpass filtered between 0.5 Hz and 100 Hz, plus an additional 50 Hz notch filter was enabled to suppress line noise. As the first 22 channels are specified for EEG, we eliminated the last 3 as they were meant to be used for monopolar electrooculography (EOG) related to visual studies Fig. 1.

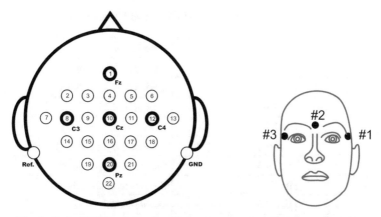

Fig. 1. Left: 22 EEG electrodes. Right: 3 monopolar EOG channels [16].

The time duration for 1 trial is in the range of 7–8 s, and there was a total of 9 runs, and each run had 48 trials, ranging from 336–384 s, with a total of 2 sessions, each one had 6 runs, 288 trials ranging 2016 and 2304 s. The first run is 2 min with eyes open, the second run is 1 min with eyes closed, and the third is 1 min with eye movements. The form of each motor imagery task is like the figure below Fig. 2.

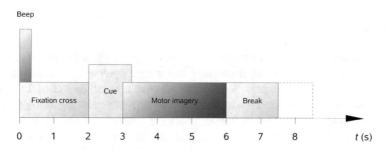

Fig. 2. Timing scheme of the MI task [16].

The data files are generally in the General Data Format for biomedical signals (GDF), one file per subject and session. This format can be loaded in Octave or MATLAB, but as this study is implemented using Python, an equivalent to the latter has been used, the same set but in numpy format (NPZ). Each subject has a list of event types; the considered are 768 at the start of the trial, 769 cue onset left (class 1), 770 cue onset right (class 2), 771 cue onset foot (class 3), and 772 cue onset tongue (class 4) and lastly 783 refers to an unknown cue, specified in the below Table 1.

Table 1. List of Subject-Related Event Types [16]

Event type		Description
276	0x0114	Idling EEG (eyes open)
277	0x0115	Idling EEG (eyes closed)
768	0x0300	Start of a trial
769	0x0301	Cue onset left (class 1)
770	0x0302	Cue onset right (class 2)
771	0x0303	Cue onset foot (class 3)
772	0x0304	Cue onset tongue (class 4)
783	0x030F	Cue unknown
1023	0x03FF	Rejected trial
1072	0x0430	Eye movements
32766	0x7FFE	Start of a new run

There are four essential keys inside each subject data, 's' contains continuous time-series recorded EEG signals. 'etype' stands for event type, which indicates event occurrence. 'epos' stands for event position. 'edur' stands for the event duration.

4.2 Design

This section intends to explore the proposed system's architecture and experimental design, combining emotion recognition and motor imagery preprocessing, feature extraction and classification techniques. The custom pipeline adopted uses ER-EEG and MI-EEG correspondingly. There are a total of 12 Jupyter Notebooks regarding this study. Each one is applied to the 9 subjects, two binary classes (left and right) are considered, and the related feature sets and classifiers are involved from both [1] and [15]. The paradigm starts by loading the BCI IV 2a dataset and preparing it for investigations by removing the last 3 EOG electrodes from the data and implementing an optimised set. As the experimentation includes all 9 subjects; however, only the usage of left and right classes (binary analysis) is included. Preprocessing filters from the emotion recognition paper related to our study [1] are being embraced. Respectively, notch filtering is applied, considering the sampling frequency Fs, the powerline frequency $powerline\_freq$ and the quality factor Q. The high-pass Butterworth filtering proceeds, considering the sampling frequency Fs, and the cutoff frequency $cutoff\_freq$. The Butterworth filter parameters consider the range (from 1 to 8), the btype (highpass) and the analogising of the signal. Downsampling is the third preprocessing phase, considering our original sampling frequency and our target one. Then, the common average reference (CAR) montage computes the mean of the EEG data across samples for each channel and ensures the result can be broadcasted and subtracted from the original EEG signal. ER-EEG feature extraction [1] is then performed in separate notebooks, starting with statistical features (mean, variance, skewness, and kurtosis), wavelet analysis (energy, entropy, and variance), higher order spectra (skewness and kurtosis), Hjorth (activity, mobility, and complexity), three-dimensional fractal dimension (Katz, Petrosian and Higuchi) and a combination of fractal dimension, wavelet analysis and Hjorth simultaneously Fig. 3.

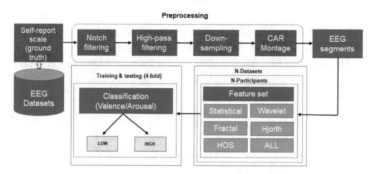

Fig. 3. ER-EEG Pipeline [1].

Followed by the MI-EEG features [15], we do not use the band-pass filters as a preprocessing filter, only the spatial filter. Applying the common spatial pattern (CSP) procedures, covariance and composite covariance, white matrix, and common eigenvector for both binary classes left and right. Features selection proceeds using mutual information-based best individual features and mutual information-based rough set theory Fig. 4.

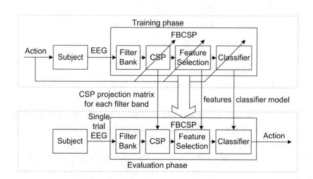

Fig. 4. MI-EEG Pipeline [15].

Generally, as classifiers, we do not use general support vector machine (SVM) [15]. However, we use the Gaussian radial basis function RBF SVM (GSVM) and the regression tree (CART) [1], and the outcome represents the train and test accuracy for all 9 related subjects, left and right classes.

4.3 Implementation

This section will explain how the previous design has been implemented during our study using the pseudocode and its algorithmic equivalent. 12 Jupyter Notebooks in total; 6 different feature sets with 2 classifiers each.

Algorithm: ER-EEG/MI-EEG Signal Processing and Classification
Input: MI EEG Dataset Consisting of Multiple Subjects
Output: Trained GSVM/CART Model and Performance Evaluation Metrics

Start Procedure:

1. Initialize Global General Parameters
2. Load BCI IV 2a Dataset
3. Dataset Preparation (Remove Unnecessary Channels)
4. For Each Subject in the Dataset:
   4.1 Preprocessing: Notch Filtering (definition and 9-subject application),
   4.2 Preprocessing: High Pass Filtering (definition and 9-subject application),
   4.3 Preprocessing: Down Sampling (definition and 9-subject application),
   4.4. Preprocessing: CAR Montage (definition and 9-subject application).
5. Feature Extraction Use Cases: (definition and 9-subject application; *separately*)
   5.1 Feature Set: Statistical Features
   5.2 Feature Set: Wavelet Analysis
   5.3 Feature Set: Hjorth
   5.4 Feature Set: Higher Order Spectra
   5.5 Feature Set: Fractal Dimension
   5.6 Feature Set: Combined (Fractal, Wavelet & Hjorth)
6. Global MI Adaptation: (definition and 9-subject application)
   6.1. Common Spatial Pattern
      6.1.1 Covariance & Composite Covariance
      6.1.2 White Matrix
      6.1.3 Common Eigenvector
   6.2. Spatial Filter
   6.3. Mutual Based Information to Select Most Informative Band
7. For Each Subject in The Dataset: (Classifiers Training/Testing Phase; *separately*)
   7.1 Train the GSVM/CART Model using The Subject's Data
   7.2 Evaluate the Trained Model using Cross-Validation on Test Data
   7.3 Store Evaluation Metrics and Test Scores
8. For Each Subject in The Dataset: (Visualize Performance)
   8.1 Plot Train Accuracy
   8.2 Plot Test Accuracy

End Procedure.

5 Experiments and Results

This section describes the experimental setup used, the evaluation metrics and gives a detailed representation of the results achieved.

5.1 Experimental Setup

The advised custom pipeline approached during this study was employed using a Mac-Book Air with the M1 Arm Apple Silicon Chip and a 16 GB of RAM. Codebase was implemented in Python 3.9.16 using the integrated development environment (IDE) Visual Studio Code 1.84.1, which is commonly used source code editor, it has local build automation, version control system integration and debuggers. The interactive computational environment for the Python code used is Jupyter Notebook, it provides an interactive coding environment, support multiple languages (in our use case: Python), it has rich text elements (markdown, LaTeX) and data visualization (Matplotlib, Plotly libraries). To containerize our workflow regarding this brain-computer interface (BCI) study, the testing and training procedures, Conda was used, which is a package management system and an environment management system that installs, runs, and updates packages and their dependencies. The essential packages being used are the following: numpy (EEG signals and multidimensional arrays), panda (manipulating numerical tables and time series), matplotlib (graph, charts visualization), scipy (high-level mathematical functions to operate on the numpy arrays) and sklearn (unsupervised and supervised machine learning algorithms).

5.2 Evaluation Metrics

The metrics used to evaluate this study's results is the Cross-Validation accuracy score [27]. The latter is a statistical method used to estimate the skill of machine learning models. The mathematical formula for the k-fold cross-validation depends on the way the average performance across all folds is computed. Let's assume that we have a dataset D, which we split into k folds. We denote each fold as D_i where i ranges from 1 to k. For each fold D_i, we train the model on $D\backslash D_i$ (the dataset D excluding the fold D_i) and then test it on D_i to get a performance score S_i. It can be given generally by CV function below (28):

$$CV = \frac{1}{k}\sum_{i=1}^{k} S_i \qquad (28)$$

where: CV is the cross-validation score, k is the number of folds and S_i is the performance score obtained when using the fold D_i as the test set and the remaining $k - 1$ fold as training set. The performance score S_i can be accuracy (being used in this study), precision, recall, F1 score, mean squared… etc.

5.3 Results

As this study had a decent number of parameters that has been fine-tuned. The showcasing of the outcome was achieved after the latter has been optimized. Firstly, the parameters overview that have been implemented and tweaked are shown in the below Table 2:

Table 2. Global Parameters Fine Tuned

Parameter	Related To	Value
notch_filter_fs	Notch Filter	250
notch_filter_powerline_freq	Notch Filter	50
notch_filter_q	Notch Filter	20.0
high_pass_filter_fs	High Pass Butterworth Filter	250
high_pass_cutoff	High Pass Butterworth Filter	0.4
down_sampling_original_fs	Downsampling Filter	250
down_sampling_target_fs	Downsampling Filter	128
car_montage_filter_axis	CAR Montage Filter	1
car_montage_filter_keep_dims	CAR Montage Filter	True
wavelet_value	Wavelet Analysis	sym3
wavelet_level	Wavelet Analysis	2
hos_nperseg	Higher Order Spectra	1024
hos_noverlap	Higher Order Spectra	512
fd_higuchi_k_max_value	Fractal Dimension: Higuchi	10
gsvm_c	GSVM Classifier	1.0
gsvm_class_weight	GSVM Classifier	None
gsvm_degree	GSVM Classifier	3
gsvm_gamma	GSVM Classifier	scale
gsvm_kernel	GSVM Classifier	rbf
gsvm_tol	GSVM Classifier	0.001
gsvm_decision_function_shape	GSVM Classifier	ovr
cart_max_depth	CART Classifier	5
cart_min_samples_split	CART Classifier	10
cart_min_samples_leaf	CART Classifier	4
cart_max_features	CART Classifier	None
cart_criterion	CART Classifier	gini
cart_splitter	CART Classifier	best
train_test_split_value	Dataset Train/Test Splitting	0.7
select_k_best_value	Mutual Based Information	3
white_matrix_m_filter	Common Spatial Pattern	2
cross_val_score_cv_value	Cross Validation Score	4

The parameters fine tuning process, started by the preprocessing filters; in our use case the Notch filter, the High Pass Butterworth filter, the downsampling filter and the CAR montage filter. Followed by the feature sets that can be optimized; in our use case the Wavelet Analysis feature extraction, the Higher Order Spectra (HOS) feature extraction and the Fractal Dimension Higuchi feature extraction. Afterwards, both accuracy performance classifiers; GSVM and CART have been adjusted. Lastly, the EEG data set Train/Test split size, the best K value for the mutual based information features, the white matrix used in the common spatial pattern (CSP) process and the cross validation (CV) value.

Secondly, the results of the used features and classifiers from the novel mixed ER-EEG and MI-EEG pipeline are given by the following accuracy Table 3:

Table 3. Results of the Embraced Pipeline

Subject		1	2	3	4	5	6	7	8	9	Mean Accuracy
Statistical	**GSVM**	56,82	59,09	56,82	52,27	56,82	52,27	72,73	77,27	63,64	**60,86**
	CART	52,27	61,36	54,55	56,82	59,09	47,73	63,64	77,27	59,09	59,09
Wavelet	GSVM	61,36	56,82	61,36	45,45	54,55	68,18	63,64	68,18	61,36	60,1
	CART	40,91	61,36	68,18	47,73	45,45	61,36	70,45	65,91	52,27	57,07
Higher Order Spectra	GSVM	56,82	47,73	52,27	54,55	52,27	50	52,27	68,18	63,64	55,3
	CART	54,55	52,27	59,09	52,27	45,45	47,73	54,55	61,36	65,91	54,8
Hjorth	GSVM	56,82	54,55	54,55	50	47,73	61,36	75	75	59,09	59,34
	CART	52,27	56,82	52,27	52,27	43,18	61,36	63,64	72,73	59,09	57,07
Fractal Dimension	GSVM	61,36	61,36	38,64	56,82	54,55	50	34,09	52,27	59,09	52,02
	CART	56,82	61,36	38,64	59,09	50	56,82	40,91	50	65,91	**53,28**
Combined 3	GSVM	54,55	50	50	61,36	47,73	47,73	45,45	54,55	56,82	**52,02**
	CART	56,82	47,73	47,73	63,64	50	52,27	50	52,27	47,73	**52,02**

Given the below chart, Fig. 5, for a visual representation of the achieved results:

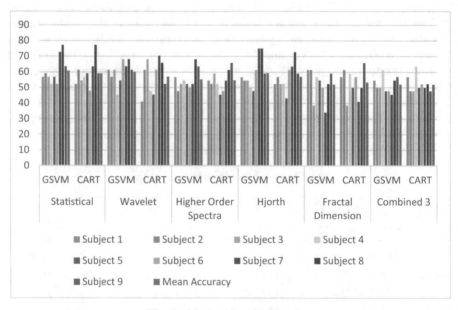

Fig. 5. Obtained Results Overview

6 Discussion

This section will interpret the results obtained, defining any related limitations and perform a comparison with the previous work from the ER-EEG scenario [1] and the MI-EEG scenario [15].

The research presented in this study offers insightful advancements in the field of EEG emotion recognition and motor imagery task recognition. By presenting noteworthy findings and comparisons with existing related studies. The results obtained shows the effectiveness of the GSVM (Generalized Support Vector Machine) classification method, which indicate higher accuracy in feature extraction, specifically when applying Statistical features. Particularly, the GSVM classifier consistently outperforms the overall results across a variety of feature sets, including Statistical, Wavelet Analysis, Higher Order Spectra (HOS), and Hjorth parameters. In contrast, the CART (Classification and Regression Trees) classifier shows higher accuracy notably with Fractal Dimension features, such as Katz, Petrosian, and Higuchi. Interestingly, a three-dimensional combination of Wavelet Analysis, Hjorth, and Fractal Dimension produces an identical mean accuracy score for both GSVM and CART classifiers, which suggests a probable convergence in classifier performance and accuracy under these considered features.

The current study is mainly scoped and limited to binary classification of left and right motor imagery tasks, using the BCI Competition IV 2a dataset. Remarkably, it worth mentioning that this research does not use two additional classes, namely tongue and feet motor imagery tasks. Which can be used in future research grounds, aiming for multi-classifying the latter. This is significant, as by integrating these additional classes, the outcome can possibly provide a more comprehensive understanding of EEG-based task classification; in our use case ER-EEG and MI-EEG.

Furthermore, this study's findings align with those presented in a different related emotion recognition article [28], which focuses on the comparison of different feature extraction methods for EEG-based emotion recognition. The latter supports the superiority of performance, precisely using the Statistical features over other feature sets such as Power, Entropy, Fractal Dimension, and Wavelet Energy, using various classification performance measures including Support Vector Machine, K-nearest Neighbors, and Decision Tree classifiers.

The research under discussion contributes novel outcome notices, particularly underlining the higher accuracy results of Statistical features over Fractal Dimension (Katz, Petrosian and Higuchi) features. It also underlines the enhanced performance of the GSVM classifier compared to the CART classifier, distinguishing this study from related literature and the implemented custom pipeline. This distinction highlights this study's unique contributions and underlines how EEG-based recognition techniques are evolving in nature, whether they are emotion related or motor tasks related. This can guide the way for future investigations and explorations in this domain of research.

7 Conclusion

Based on the findings from this study, it is evident that the utilization of feature extraction methods traditionally applied in emotion recognition, when employed in the arena of motor imagery, produces distinct outcomes. The research highlights that binary classification can effectively be used to analyze EEG data for both emotion recognition (valence and arousal) and motor imagery (left and right classes). Notably, statistical feature extraction methods, particularly compared to the measuring of fractal dimensions methods, delivers superior accuracy in motor imagery tasks. As referred in the discussion section, the related article underlines this study's findings. Moreover, when comparing classifier performance on the BCI Competition IV 2a dataset, the Gaussian radial basis function SVM (GSVM) consistently outperforms the regression tree (CART). These findings are important as they suggest that the cross-application of methods between different areas of EEG analysis can yield to unique results, and that certain methods may be more beneficial depending on the specific application within the study field of EEG-related emotion recognition and/or motor imagery tasks.

References

1. Yuvaraj, R., Thagavel, P., Thomas, J., Fogarty, J., Ali, F.: Comprehensive analysis of feature extraction methods for emotion recognition from multichannel EEG recordings. Sensors **23**(2), 915 (2023). https://doi.org/10.3390/s23020915
2. Soleymani, M., Lichtenauer, J., Pun, T., Pantic, M.: A multimodal database for affect recognition and implicit tagging. IEEE Trans. Affect. Comput. **3**(1), 42–55 (2012)

3. Koelstra, S., et al.: DEAP: a database for emotion analysis using physiological signals. IEEE Trans. Affect. Comput. **3**(1), 18–31 (2012)
4. Zheng, W.L., Lu, B.L.: Investigating critical frequency bands and channels for EEG-based emotion recognition with deep neural networks. IEEE Trans. Auton. Ment. Dev. **7**(3), 162–175 (2015)
5. Miranda-Correa, J.A., Abadi, M.K., Sebe, N., Patras, I.: AMIGOS: a dataset for affect, personality, and mood research on individuals and groups. IEEE Trans. Affect. Comput. **12**(4), 479–493 (2017)
6. Katsigiannis, S., Ramzan, N.: DREAMER: a database for emotion recognition through EEG and ECG signals from wireless low-cost off-the-shelf devices. IEEE J. Biomed. Health Inform. **22**(1), 98–107 (2018)
7. Jenke, R., Peer, A., Buss, M.: Feature extraction and selection for emotion recognition from EEG. IEEE Trans. Affect. Comput. **5**(3), 327–339 (2014)
8. Liu, Y., Sourina, O.: Real-time subject-dependent EEG-based emotion recognition algorithm. In: Gavrilova, M.L., Kenneth Tan, C.J., Mao, X., Hong, L. (eds.) Transactions on Computational Science XXIII: Special Issue on Cyberworlds, pp. 199–223. Springer Berlin Heidelberg, Berlin, Heidelberg (2014). https://doi.org/10.1007/978-3-662-43790-2_11
9. Yuvaraj, R., et al.: Optimal set of EEG features for emotional state classification and trajectory visualization in Parkinson's disease. Int. J. Psychophysiol. **94**(3), 482–495 (2014)
10. Nawaz, R., Cheah, K.H., Nisar, H., Yap, V.V.: Comparison of different feature extraction methods for EEG-based emotion recognition. Biocybern. Biomed. Eng. **40**(4), 910–926 (2020)
11. Liu, J., Meng, H., Li, M., Zhang, F., Qin, R., Nandi, A.K.: Emotion detection from EEG recordings based on supervised and unsupervised dimension reduction. Concurrency Comput. Pract. Exp. **30**, e4466 (2018)
12. Katz, M.J.: Fractals and the analysis of waveforms. Comput. Biol. Med. **18**(3), 145–156 (1998)
13. Hatamikia, S., Nasrabadi, A.M.: Recognition of emotional states induced by music videos based on nonlinear feature extraction and SOM classification. In: Proceedings of the 21st Iranian Conference on Biomedical Engineering (ICBME), Tehran, Iran, pp. 333–337 (2014)
14. Higuchi, T.: Approach to an irregular time series on the basis of the fractal theory. Physica D **31**(1–2), 277–283 (1988). https://doi.org/10.1016/0167-2789(88)90081-4
15. Ang, K.K., Chin, Z.Y., Wang, C., Guan, C., Zhang, H.: Filter bank common spatial pattern algorithm on BCI competition IV datasets 2a and 2b. Front. Neurosci. **6**, 39 (2012). https://doi.org/10.3389/fnins.2012.00039
16. Schlögl, A., et al.: BCI Competition 2008 – Graz data set A. Institute for Knowledge Discovery (Laboratory of Brain-Computer Interfaces), Graz University of Technology (2008). http://www.bbci.de/competition/iv/
17. Wang, X.W., Nie, D., Lu, B.L.: Emotional state classification from EEG data using machine learning approach. Neurocomputing **129**, 94–106 (2014)
18. Hjorth, B.: EEG analysis based on time domain properties. Electroencephalogr. Clin. Neurophysiol. **29**(3), 306–310 (1970)
19. Hjorth, B.: The physical significance of time domain descriptors in EEG analysis. Electroencephalogr. Clin. Neurophysiol. **34**(3), 321–325 (1973)
20. Hosseini, S.A.: Classification of brain activity in emotional states using HOS analysis. Int. J. Image Graph. Sig. Process. **4**(1), 21 (2012)
21. Hirano, K., Nishimura, S., Mitra, S.: Design of digital notch filters. IEEE Trans. Commun. **22**(7), 964–970 (1974)
22. Hussin, S.F., Birasamy, G., Hamid, Z.: Design of butterworth band-pass filter. Politeknik Kolej Komuniti J. Eng. Technol. **1**(1) (2016)

23. Youssef, A.: Image downsampling and upsampling methods. National Institute of Standards and Technology (1999)
24. Lemos, M.S., Fisch, B.J.: The weighted average reference montage. Electroencephalogr. Clin. Neurophysiol. **79**(5), 361–370 (1991)
25. Yang, J., et al.: Parameter selection of Gaussian kernel SVM based on local density of training set. Inverse Prob. Sci. Eng. **29**(4), 536–548 (2021)
26. Azuaje, F.: Witten IH, Frank E: data mining: practical machine learning tools and techniques. Biomed. Eng. Online 5, **51** (2006)
27. Diamantidis, N.A., Karlis, D., Giakoumakis, E.A.: Unsupervised stratification of cross-validation for accuracy estimation. Artif. Intell. **116**(1–2), 1–16 (2000)
28. Nawaz, R., Cheah, K.H., Nisar, H., Yap, V.V.: Comparison of different feature extraction methods for EEG-based emotion recognition. Biocybern. Biomed. Eng. **40**(3), 910–926 (2020)

Control of Respiratory Ventilators Using Boussignac Valve

Zbigniew Szkulmowski[1], Sławomir Grzelak[2], Michał Joachimiak[2], Sebastian Meszyński[2(✉)], Marcin Schiller[3], and Oleksandr Sokolov[2]

[1] Home Ventilation Center for Specialized Medical Care, Bydgoszcz, Poland
[2] Nicolaus Copernicus University, Toruń, Poland
sem@umk.pl
[3] Antoni Jurasz University Hospital No. 1, Bydgoszcz, Poland

Abstract. The aim of this study was modelling and computer simulation of the ventilator control system in the pressure-controlled ventilation (PCV) model with the use of the Boussignac valve and the construction of a prototype of such a ventilator. The simulation model was based on Matlab/Simulink model of Medical Ventilator with Lung Model. This model is a nonlinear system with a fuzzy Takagi-Sugeno rule base model. A program interface was developed for simulating the operation of ventilator in the PCV mode on-line and to change the mechanical parameters of the lung - resistance and compliance.

Keywords: lung simulation · Boussignac · control · fuzzy logic

1 Introduction

It is well known that the Covid epidemic in 2020–2022 was a challenge not only for the patients over the world, but especially for medical personnel, doctors and medical facilities. Characteristic features of coronavirus infection are its low controllability and the possibility to develop pneumonia in the course of infection, in some cases with severe course, with respiratory failure requiring the use of mechanical ventilation. A ventilator is a medical device that supports or replaces the function of patient muscles in breathing work. The machine provides artificial, forced respiration, used in case of respiratory function cessation, as a result of injury, disease or the neuromuscular blocking drugs usage. It is also used to facilitate breathing in cases when the patient's respiratory system failure or its efficiency does not meet the body's need for oxygen).

There are two major modes of ventilations using respirators (see Fig. 1) [1]:

- volume-controlled ventilation (VCV),
- pressure-controlled ventilation (PCV)

In general, volume control favors the control of ventilation based on flow dV/dt, and pressure control favors the control of oxygenation based on pressure P. Volume and pressure control modes have distinct advantages and disadvantages which are mainly related to the flow and pressure patterns of gas delivery. Volumetric supply is a gas reservoir of a certain volume, from which the inspiratory air flow is forced with each inhalation towards the patient's lungs. Pressure supply is a hospital installation or turbine that allows the supply of gas with high, variable and unrestricted flows and high pressures with a fast rise time.

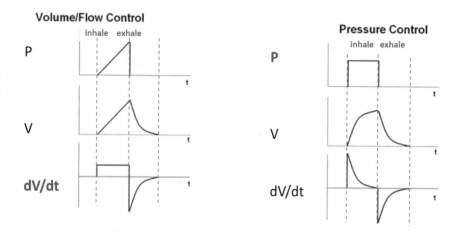

Fig. 1. Ventilation control modes; a) volume-controlled ventilation; b) pressure-controlled ventilation

The control of ventilators determines their capabilities in terms of supporting the patient's own breathing. These possibilities are greater with PCV. This is due to the ability of these respirators to provide a wide range of gas flow, and especially high instantaneous flow (see Fig. 1). This allows the inspiratory flow to be adjusted to match natural flow fluctuations during spontaneous and ventilator-assisted breaths and maintains a constant inspiratory pressure in the event of strenuous breathing and in the event of inadvertent air leakage between the face mask and the patient's face during non-invasive ventilation.

The PCV also has other advantages over a VCV that are of great importance when treating respiratory failure associated with viral pneumonia. The article describes ventilator control in PCV mode. A conventional ventilator is controlled by valves at the inlet and outlet of the air stream. Constant care and sterilisation of mechanical valves is too burdensome, especially during epidemics and frequent use of respirators. Therefore, the use of the Boussignac valve in the exit channel was designed. In addition, only one Boussignac valve at the outlet was proposed to control the ventilator, instead of the traditional two-valve scheme.

The aim of this study was modelling and computer simulation of the ventilator control system in the PCV mode with the use of the Boussignac valve and the construction of a prototype of such a ventilator.

2 Model of Respirator-Human Interaction Using Boussignac Valve Simulator

2.1 Model of the Boussignac Valve and Control System

Boussignac CPAP (Vygon (UK) Ltd) is a device integrated into the respiratory support system, shaped in the form of cylinder positioned between the patient's mask and the external environment. It has been designed for patients with spontaneous ventilation, allowing the maintenance of higher airway pressure throughout the entire breathing cycle compared to atmospheric pressure [2]. The elevated pressure is generated using the phenomenon of turbulence. This turbulence arises as gases administered to the patient pass through 4 microchannels, accelerating gas particles significantly, and inject them into the chamber. Inside this chamber, due to the turbulence, the particles collide, acting as the virtual valve [3, 4]. This device includes 2 additional ports: the first one allows connection to an external oxygen source, and the second port enables the attachment of an additional oxygen source, permitting pressure control or CO_2 monitoring (see Fig. 2).

In clinical practice, CPAP significantly improves the treatment outcomes in various respiratory failure conditions such as acute cardiogenic pulmonary edema, pneumonia associated with immune deficiencies, or postoperative complications. Thanks to its effectiveness in reducing breathlessness, improving gas exchange, and vital parameters, it often limits the need for endotracheal intubation. The Boussignac CPAP device is particularly significant in the management of respiratory failure, offering a lightweight, portable, and user-friendly system that proves effective even for patients withdrawn from respiratory support. It is also cheap and easy to exchange in case of high demand for antiseptic solutions.

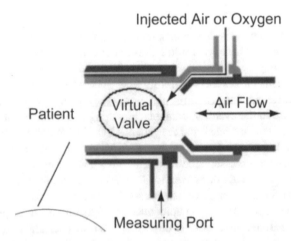

Fig. 2. The Boussignac continuous positive airway pressure system after [3]

Its advantages include the absence of moving parts, eliminating the risk of blockages, and an open design that allows patients to breathe freely even in the event of interrupted oxygen supply from the bottle [5].

The common disadvantage found in setups that utilize Boussignac valve is the correlation between PEEP level and inspired oxygen fraction (F_iO_2) and thus, lack of independent regulation for these parameters [6]. In addition, high F_iO_2 and PEEP more than 20 cmH_2O are not possible to achieve [7, 8].

As mentioned above, the Boussignac valve works by creating a turbulent vortex when air flows through the Injected Air channel, which blocks the main outlet channel from the patient's mask. To simulate this effect and design the control system we propose to use simple rules like follow:

if Injected Air Flow is big then Boussignac valve is closed,

if Injected Air Flow is middle, then Boussignac valve is semi closed,

if Injected Air Flow is zero, then Boussignac valve is open.

In order to add smoothness to the modelling and control process, we suggest using fuzzy logic controller, which enables smooth switching between the rules (see Fig. 3). Fuzzy model is based on Sugeno 0 type interface and was created using ANFIS toolbox in Matlab [9].

The model of mechanics of ventilation consists of a mechanical model of lung [10], model of gas transportation [12], approximation of the Boussignac valve [3]. Part of inlet flow goes through the Boussignac valve that is controlled by an additional proportional valve. The scheme of flows in the respiration system using Boussignac valve is shown in Fig. 4.

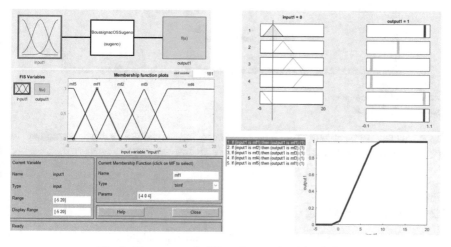

Fig. 3. Fuzzy control of Boussignac valve (simulation)

To simulate the Boussignac valve as a nonlinear system the fuzzy Takagi-Sugeno rule base model is proposed [11]. It transforms the input value of the air flow into percent of valve state (1 - it is open, 0 - it is closed).

Each rule generates one value:

$$Output_i = a_i \cdot x + b_i \tag{1}$$

where ai and bi are constant coefficients, x is an input signal.

Rule firing strength w_i derived from the rule antecedent

$$w_i = F_i(Input) \tag{2}$$

here, $F_i(x)$ are the membership functions for input x.

The output of each rule is the weighted output level, which is the product of w_i and Output$_i$.

The final output of the system is the weighted average over all rule outputs:

$$FinalOutput = \frac{\sum_{i=1}^{N} w_i \cdot Output}{\sum_{i=1}^{N} w_i} \tag{3}$$

The fuzzy model will be useful for approximation of real characteristics of Boussignac valve for instance using neuro-fuzzy Inference algorithm (ANFIS).

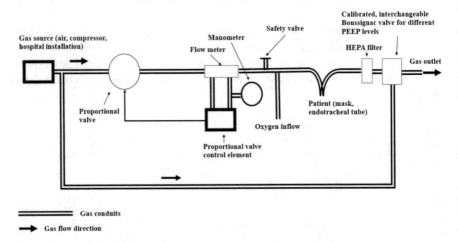

Fig. 4. Scheme of ventilator with lung mechanics and Boussignac valve - controlling the proportional valve using data from the flow meter and pressure sensor

2.2 Compartment Model of Lung

In general, a compartment model is a type of mathematical model used for describing the way materials or energies are transmitted among the compartments of a system. In the case of lung mechanics in invasive case (without the leaks) the following model is used:

$$P_{vent} + P_{mus} = \frac{V_T}{C_{RS}} + R_{aw} \times \dot{V}_i + PEEP + PEEPi$$

where P_{vent} is the airway pressure applied by the ventilator, P_{mus} is the pressure generated by the patient's inspiratory muscles, V_T is the tidal volume, C_{RS} is the compliance of the respiratory system, R_{aw} is resistance in the respiratory tract, dV_i/dt is the inspiratory flow, PEEP to PEEP (positive end expiratory pressure) set on the ventilator, PEEPi is self-peeling (auto-PEEP).

Here resistance R is a measure of the resistance to gas flow caused by the internal friction of the breathing air flow, as well as the friction between the breathing air and the airway ($cmH_2O/l/s$). Compliance C is the lung extensibility index, i.e., the ability of the lungs to expand, defined as the change in lung volume (ΔV) per unit change in pressure (ΔP). Unit: ml/cmH_2O. Using model (1) one can simulate the breathing process (see Fig. 5) using different control methods.

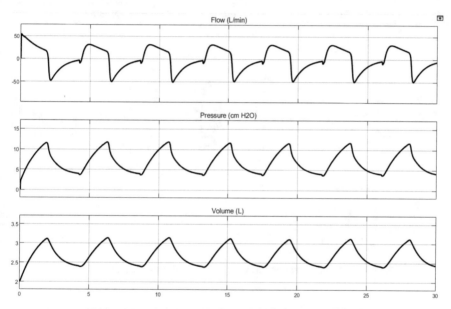

Fig. 5. Example of simulation of breathing process in time (s)

2.3   System Control Based on Boussignac Valve

There are several levels of control. The bottom level is a base program performing, the top level is a decision of medical personnel about changing the parameters of respirators. There are many types of respiration control in the bottom level. They depend on the medical decision depending on if it will be controlled by mandatory ventilation or assist control. In the last case the muscle activity during inhalation should be taken into account where the trigger mechanism is used. We considered several types and built-up control schemes for pressure-controlled ventilation.

System of control based on the Boussignac valve is built using two proportional valves P1, P2 and Boussignac valve (B). In the phase of inspiration B should be closed, that means that P1 should be open. It causes the turbulence effect in the Boussignac valve and blocks the expiration channel. This means that in order to increase the flow rate during the inspiratory phase, valve P2 must also be open. It is also obvious that in the exhalation phase the valve P2 must be closed. The following is a table of valve limit positions in both phases (see Fig. 6 and Fig. 7).

Table 1. Valve Limit Positions in Inspiratory and Exhalation Phases for the Boussignac Valve-Based Control System

P1	P2	B	Phase
open	open	close	inspiration
close	close	open	expiration

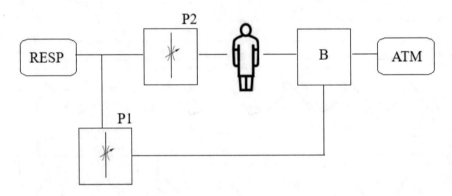

Fig. 6. Illustration depicting the valve positions during the inspiratory phase

Taking into account the different diameters of the tubes leading from the compressor to the patient and the Boussignac valves (smaller into Boussignac valve), as well as the switching logic Table 1, it is possible to use only one proportional valve. In this case, a modified scheme may be proposed.

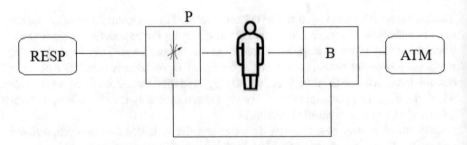

Fig. 7. Visualization of valve positions during the exhalation phase

Next figure illustrates the scheme and simulation of the control system with the Boussignac valve (see Fig. 8 and Fig. 9). The model was built using SimScape toolbox, that allows to use moist air blocks and process different physical signals using pressure and flow sensors [13].

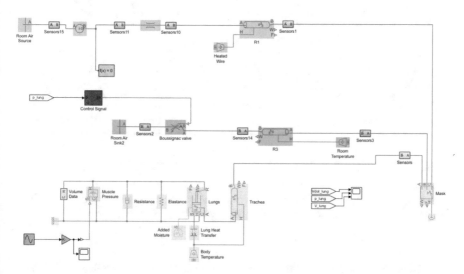

Fig. 8. The scheme of the respirator control system with the Boussignac valve (yellow – model of lung with mask; grey – respirator with pressure control; blue – out channel with Boussignac valve; red – block with Boussignac fuzzy control; green – simulator of muscle activities)

For the purpose of simulating the operation of the ventilator and the lung model, a program interface was developed (see Fig. 10). It can be used to manually control the parameters of the ventilator in the PCV mode on-line and to change the mechanical parameters of the lung - resistance and compliance. The interface was developed using Dashboard blocks [14] that connect with subsystem of breathing model (see Fig. 8) using wireless connection mechanism.

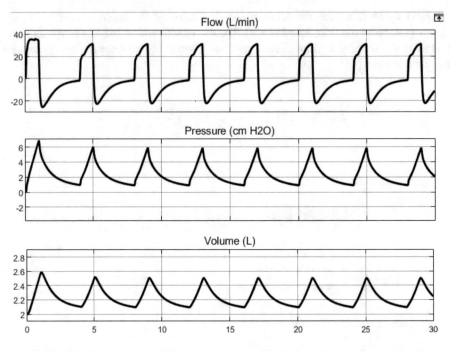

Fig. 9. Simulation results of the control system with the Boussignac valve in time (s)

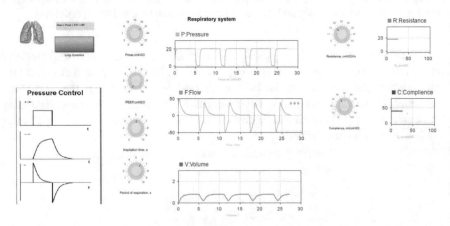

Fig. 10. Interface of respirator-lung model and manual control according to regular respirators. Left side control buttons are standard buttons of many respirators with Pmax, PEEP, respiration time and period of respiration. Right side buttons serve to change the mechanical model of lung – resistance and compliance

3 Conclusion

The article presents an interaction model between the respirator and the human body. A fuzzy model for the Boussignac valve was proposed, and control algorithms were developed for various breathing patterns using Boussignac valves. Through simulations, it was demonstrated that a single Boussignac valve is sufficient for effective control. Future work will focus on the integration of different control modes into a unified system, advancing the field of respiratory control with the Boussignac valve. This study serves as a foundational step towards a comprehensive understanding of this control system.

Future research related to the Boussignac valve, and its application will encompass two areas aimed at acquiring knowledge for a better understanding of the device's mechanisms. This will lead to the refinement and optimization of such devices. This new knowledge may contribute to improving the effectiveness and efficiency of the Boussignac valve. As a result, better therapeutic outcomes can be achieved in clinical application. Below are the two areas of research:

- Numerical simulations and computer modeling - conducting advanced numerical simulations for the Boussignac valve under various gas flow conditions. Utilizing computer modeling techniques such as finite element method to analyze the dynamics of gas flow through the valve. Research may focus on optimizing the valve's geometry to enhance its efficiency.
- Laboratory experiments using sensors - conducting laboratory experiments with the use of sensors to analyze airflow parameters of air or oxygen through the Boussignac valve. These studies may involve measurements of pressure, flow velocity, gas volume changes during interaction with the valve.

References

1. West, J.B., Luks, A.M.: West's Respiratory Physiology. Lippincott Williams & Wilkins (2020)
2. Sehlin, M., Törnell, S.S., Öhberg, F., Johansson, G., Winsö, O.: Pneumatic performance of the Boussignac CPAP system in healthy humans. Respir. Care **56**, 818–826 (2011)
3. Mistraletti, G., et al.: Noninvasive CPAP with face mask: comparison among new air-entrainment masks and the Boussignac valve. Respir. Care **58**, 305–312 (2013)
4. Volsko, T.A.: Devices used for CPAP delivery. Respir. Care **64**, 723–734 (2019)
5. Dieperink, W., Goorhuis, J.F., de Weerd, W., Hazenberg, A., Zijlstra, J.G., Nijsten, M.W.N.: Walking with continuous positive airway pressure. Eur. Respir. J. **27**, 853–855 (2006)
6. Bellani, G., Foti, G., Spagnolli, E., Castagna, L., Patroniti, N., Pesenti, A.: An improved Boussignac device for the delivery of non-invasive CPAP: the SUPER-Boussignac. Intensive Care Med. **35**, 1094–1099 (2009)
7. Wong, D.T., Tam, A.D., Van Zundert, T.C.: Others: The usage of the Boussignac continuous positive airway pressure system in acute respiratory failure. Minerva Anestesiol. **79**, 564–570 (2013)
8. Vargas, M., et al.: Performances of CPAP devices with an oronasal mask. Respir. Care **63**, 1033–1039 (2018)
9. Tamburrano, P., De Palma, P., Plummer, A.R., Distaso, E., Amirante, R.: Simulink Modelling for Simulating Intensive Care Mechanical Ventilators. In: E3S Web of Conferences, p. 7007 (2020)

10. J.H.T.: Lung mechanics: an inverse modeling approach. Cambridge University Press (2009)
11. Jang, J.-S.: ANFIS: adaptive-network-based fuzzy inference system. IEEE Trans. Syst. Man Cybern. **23**, 665–685 (1993)
12. Mathworks   Homepage,   https://www.mathworks.com/help/simscape/gas-models.html, Accessed 4 Jan 24
13. Mathworks Homepage, https://www.mathworks.com/products/simscape.html, Accessed 4 Jan 24
14. Mathworks   Homepage,   https://www.mathworks.com/help/simulink/dashboard.html, Accessed 4 Jan 24

Deep Learning Brain MRI Segmentation and 3D Reconstruction: Evaluation of Hippocampal Atrophy in Mesial Temporal Lobe Epilepsy

Aymen Chaouch[1,2]([envelope]), Nada Hadj Messaoud[1], Asma Ben Abdallah[1], Jamal Saad[3], Laurent Payen[4], Badii Hmida[3,4], and M. Hedi Bedoui[1]

[1] Medical Technology and Image Processing Laboratory, Faculty of Medicine Monastir, Monastir, Tunisia
ch.aymen@gmail.com
[2] Higher Institute of Computer Science and Communication Technologies, H.Sousse, University of Sousse, Sousse, Tunisia
[3] Medical Imaging Department, Fattouma Bourguiba Hospital, Monastir, Tunisia
[4] Medical Imaging Department, Saint-Denis Hospital, Saint-Denis, France

Abstract. In the present paper, we present an automated approach to analyse the lateralization of the epileptogenic focus within the mesial temporal lobe (mTLE) using brain MRI scans. The proposed method encompasses an initial segmentation stage that utilizes a deep convolutional neural network (CNN), followed by a 3D reconstruction and volume computation for both the right and left hippocampus. Our comprehensive approach involves preprocessing the database, employing augmentation techniques, and evaluating outcomes using standard metrics. We validated our method using a dataset containing MRI images from 50 patients, resulting in encouraging findings. Specifically, our approach achieved an 0.84 Dice score, surpassing the values previously reported for this dataset in the literature. Moreover, our method attained an 89% sensitivity and a Hausdorff distance of 2.59.

Keywords: Epilepsy · Hippocampus · Deep Learning · Segmentation · Preprocessing MRI · 3D reconstruction

1 Introduction

Epilepsy is a chronic neurological disorder affecting numerous individuals worldwide. It is characterized by recurrent seizures triggered by abnormal electrical discharges in the brain, which can lead to convulsions, loss of consciousness and hallucinations (visual, olfactory, or auditory). While medication can effectively manage seizures in most cases (between 70 and 80%), surgery may be a viable option for cases of partial epilepsy (i.e., epilepsy that is restricted to a specific region of the brain) that does not respond to medication. Accurate localization of the epileptogenic zone is essential for this procedure.

© The Author(s), under exclusive license to Springer Nature Switzerland AG 2024
Á. Rocha et al. (Eds.): WorldCIST 2024, LNNS 986, pp. 243–253, 2024.
https://doi.org/10.1007/978-3-031-60218-4_22

In clinical practice, various modalities are employed for diagnosing epileptic seizures, such as EEG (Electroencephalogram), MRI (Magnetic Resonance Imaging), PET (Positron Emission Tomography). These techniques have enhanced diagnosis by enabling direct observation of phenomena that previously had to be inferred. In brain imaging specifically, each technique has distinct purposes. For instance, structural imaging (e.g., computed tomography, MRI) allows for the study of brain anatomy and is used to diagnose various conditions such as tumors, hemorrhages, degenerative brain diseases, and more. Functional imaging (e.g., PET, fMRI, MEG) maps brain activity, enabling visualization of deep brain structures, identification of epileptic foci, and detection of multiple sclerosis lesions.

All these brain imaging techniques consider complementary aspects of brain anatomy and activity. Combined with developments in information technology, they have led to a real revolution in our knowledge of how the brain works.

In this particular context, and specifically for the purpose of diagnosing the type of epilepsy under examination, different modalities are employed. The electroencephalogram (EEG) is the primary diagnostic tool for identifying epileptic seizures and their classifications. However, the EEG is unable to detect the precise location and microstructural changes related to the epileptogenic foci.

The MRI provides structural images with good spatial resolution and is particularly useful for studying the anatomy of the brain in vivo and providing information on the atrophy of its structures (the hippocampus in our case) [1, 2] Positron Emission Tomography (PET) extracts functional and metabolic information and can be used for early diagnosis of the disease [3].

In the realm of our study, focusing on mesial temporal lobe epilepsy (mTLE), research indicates that the majority of seizures can be accurately anticipated using hippocampal information [1, 2, 4]. Most patients exhibit hippocampal atrophy, marked by neuronal loss [5]. Manual hippocampal delineation stands as the benchmark, offering precision and high sensitivity in atrophy detection [1]. However, this method is laborious and time-consuming. Consequently, several studies have proposed automated hippocampal segmentation methods, particularly leveraging Machine Learning (ML) and Deep Learning (DL). These methods have aimed at diagnosing various conditions such as epilepsy, Alzheimer's, and depression [6–8]. Yet, limited attention has been devoted to automatic hippocampal segmentation specifically for mTLE [9–11].

For the lateralization of epileptogenicity in mesial temporal lobe epilepsy (mTLE) hippocampus volumetry analysis is necessary and is generally obtained by manual segmentation. However, this process is both tedious and time-consuming, hence automatic segmentation algorithms are preferred. Numerous automated segmentation techniques have been proposed. The method we are proposing involves setting up a complete automatic chain for assessing asymmetry between 3D objects reconstructed from 2D parallel image sequences segmented using a DL approach (Fig. 1).

2   Subjects and MR Imagining

Our mTLE database consists of 50 subjects supplied by Kouroush Jafari [11], 25 with ground truth and 25 without labelling. Of those labelled, 5 were found to be healthy, 10 with right lateralization (R), 8 with left lateralization (L), and 2 epileptic patients whose

Fig. 1. Workflow of the proposed segmentation and lateralization method. a complete automatic chain for assessing asymmetry between 3D objects reconstructed from 2D parallel image sequences segmented using a DL approach. A similarity index based on the difference in object volumes is calculated.

lateralization was not specified. MRI acquisitions were performed on 1.5T and 3T MRI scanners at the Henry Ford Hospital Neurology Department, Detroit, USA.

30 patients underwent coronal (T1) 1.5 T MRI (General Electric Signa, GE Medical Systems, Milwaukee, WI, USA), a deteriorated gradient echo sequence (SPGR) with the following parameters: TR = 7.6 ms, TI = 1.7 ms, TE = 500 ms, flip angle = 20°, field of view (FOV) = 200 × 200 mm². Matrix size is 256 × 256, pixel resolution is 0.781 × 0.781 mm² and slice thickness is 2.0 mm.

A total of 20 patients underwent a 3 T General Electric 3.0T MRI system (GE Medical Systems, Milwaukee WI). A deteriorated gradient echo sequence (SPGR) with the following parameters: TR = 10.4 ms, TI = 4.5 ms, TE = 300 ms, flip angle = 15°, FOV = 200 × 200 mm². The matrix size is 512 × 512, the pixel resolution is 0.39 × 0.39 with a slice thickness = 2.0 mm. The regions of interest encompassing the hippocampi were segmented manually (with an average of 19 slices per case). Table 1. Reports the database information. We note the difference in resolution between the 1.5 T and 3.0 T images (Fig. 2.)

Table 1. Data set characteristics

Subject number	Modality	MRI plan	Slice number	Labelled slices number	Voxel size	Image size
Subjects with ground truth						
15	IRM T1 (1.5 T)	Coronal	1848	276	0.78 × 0.78 × 2	256 × 256
10	IRM T1 (3 T)		1202	183	0.39 × 0.39 × 2	512 × 512
Total: 25			3050	459		
Subjects without ground truth						
15	IRM T1 (1.5 T)	Coronal	1696	276	0.78 × 0.78 × 2	256 × 256
10	IRM T1 (3 T)		1142	182	0.39 × 0.39 × 2	512 × 512
Total: 25			2838	458		
Total			5888	917		

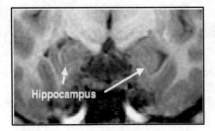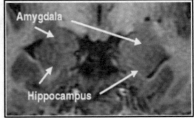

Fig. 2. From the left to the right 1.5T, 3T

3 Proposed Approach for the 2D Image Segmentation and 3D Objects Asymmetry Analysis

Our proposed approach is composed of five steps: (i) Database structuring and 2D image preprocessing (Fig. 3), (ii) Data augmentation and patch extraction, (iii) 2D image segmentation, (iv) A 3D reconstruction using segmented parallel 2D images as input and generating a triangular surface mesh describing the surface of the object as output, (v) The meshes describing the structures of interest will be used to calculate volumes and the asymmetry index.

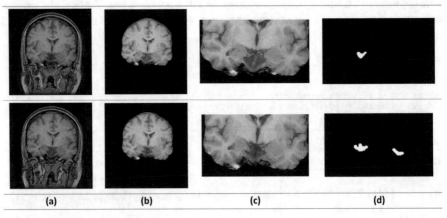

Fig. 3. (a) original image size 256 x 256, (b) Skull Striping (c), image cropping and histogram equalization, (d) Ground Truth

3.1 Preprocessing and Data Augmentation

Data Augmentation

We used coronal slices extracted from MRI volumes. As preprocessing, we applied (1) skull stripping, (2) intensity normalization, (3) histogram equalization, (4) image size standardization to 256×256 and finally (5) a cropping size of 60×190 that covers our region of interest (Fig. 2). Data augmentation techniques were implemented to increase the training data and to improve the performance of the network as well as to prevent the network from memorizing the exact details of the training dataset. Two geometric transformations were used for this study, namely rotation and vertical flip.

More specifically, the Keras Image Data Generator class was implemented with the following arguments: Rotation_range = 10, Horizontal_Flip, brightness_range = [0.2, 1.0]. This class was used to apply a random combination of these transformations to the images entered. During the testing phase, a total of 74 images were available for patients. However, during the training phase, the number of images reached a final amount of 9273.

Patch Extraction

The technique involves dividing the image into a series of patches to uphold training accuracy, increase the amount of training data and identify small structures (e.g., Hippocampus). In practice, segmentation of the hippocampus can pose challenges due to the shape alterations associated with neurological disorders, resulting in dimension and shape discrepancies. Segmentation of a whole image using deep-learning architectures yields imprecise outcomes. To enhance system performance and overcome this issue, a patch extraction method was implemented (Fig. 4). This method involves dividing the image into a set of patches with specific dimensions. In the present study, the patch size was determined empirically bearing in mind that it should exceed the size of the hippocampus structure to achieve more precise segmentation. The patch size was 32 × 32.

Fig. 4. Patch extraction examples

DL Segmentation Architectures

U-Net and ResNet networks have been widely applied in medical imaging and have shown excellent performance in segmentation [12, 13]. U-Net_v2, our architecture based on U-Net. Where we added a batch normalization and a dropout layers.

3.2 Post-processing

The segmented patches were merged and a threshold was applied to create a binary image of the hippocampus. All segmented patches within the image were collected and arranged in a slicing order based on image size. An empirical threshold was then applied to enhance the resulting image.

3.3 Hippocampus Volume and Asymmetry Index Computing

The segmented images will serve as input for the 3D reconstruction procedure. Our study involved reconstructing the object surface in 3D using a series of parallel images, which were segmented to generate a triangular mesh that represented the surface. The Marching Cubes algorithm [14] was employed for this purpose as it is a highly recognized and widely used approach in medical imaging for generating 3D models of anatomical structures from 2D serialized and segmented images.

Once the triangular mesh has been generated, the asymmetry between input objects is quantified.

This is was based on the Green-Ostrogradsky formula [15] known by various names: Gauss's theorem, or the divergence theorem Defined as follows:

The hippocampus's degree of asymmetry was determined by obtaining the percentage ratio of the absolute difference between the left and right volumes to their sum [16].

$$AI = \frac{|R - L|}{(R + L)0.5} \tag{1}$$

When the degree of asymmetry decreases in a specific structure, indicated by lower AI values, for this work the AI measures asymmetry directed towards the epileptic focus in mTLE, where AI = 0 when Left = Rights. The right hippocampus volume is denoted by R, while the left hippocampus volume is denoted by L. Results and evaluation metrics.

In this section, we describe the results obtained in the 3D object asymmetry analysis phase and its application in the case of mTLE patients. We begin by presenting the metrics used to evaluate the DL-based segmentation approaches. Next, the results of the segmentation, 3D reconstruction and analysis of the asymmetry between 3D objects will be explained and discussed.

3.4 Evaluation Metrics

The proposed method, along with three additional methods - HAMMER [10], FreeSurfer [17], two publicly available software tools, and localInfo [11], were applied in the segmentation of MR images of all hippocampi.

Before discussing the evaluation metrics presented in Table 2, it is necessary to clarify the four categories of statistical classification, known as the confusion matrix. These are outlined below:

- True positive (TP): when the model correctly predicts the positive class.
- True negative (TN): The model correctly predicts the negative class.

- False positive (FP): an outcome where the model incorrectly indicates the positive class, but it is actually negative.
- False negative (FN): an outcome where the model incorrectly indicates the negative class, but it is actually positive.

The following measures were used to evaluate and compare segmentation models.

Table 2. Evaluation metrics

Precision	Sensitivity	Dice	Hausdorff Distance
$\dfrac{TP}{TP+FP}$	$\dfrac{TP}{TP+FP}$	$\dfrac{2*TP}{2*(TP+FN+FP)}$	$hd(X,Y) \max\limits_{x \in X} \min\limits_{y \in Y} \lvert x - y \rvert_2$

3.5 DL Segmentation Results

We implemented 3 DL segmentation models: U-Net, U-Resnet, U-Net_V2 and we used the mTLE database composed of 50 patients. This database is divided into three parts: learning, validation and testing. 90% of the data was used in the learning phase and 10% for testing. The cross-validation method was used, with a percentage of 20% for validation and 80% for learning. Table 3 shows the best metric values calculated for each architecture implemented after hyperparameter optimization.

Table 3. Comparison of results obtained by DL segmentation and other automatic methods

Methods	Dice	Sensitivity	Precision	Hausdorff Distance
Deep Learning segmentation				
U-net [12]	0.82	0.77	**0.88**	2.92
U-ResNet [13]	0.78	0.72	0.85	3.03
U-Net_V2	**0.84**	**0.89**	0.77	**2.59**
Other automatic methods				
FreeSurfer [17]	0.64	0.68	x	5.89
HAMMER [10]	0.63	0.64	x	4.19
LocalInfo [11]	0.72	0.66	x	3.09

3.6 3D Reconstruction and Volume Quantification

Two parallel segmented image sequences are input to our 3D reconstruction process, each representing the right and left hippocampus of each patient. Our output generates two triangular surface meshes that describe the two surfaces of our 3D structures of interest, specifically, the right and left hippocampus. Figure 5 presents a surface rendering of the hippocampus, derived from an MRI image sequence that was previously segmented. All patients had their hippocampal volumes calculated on both sides. We compared manual segmentation with our automatic segmentation method using four randomly selected controls (HFH-003, HFH-007, HFH-010, HFH-020) during testing.

(a)	(b)

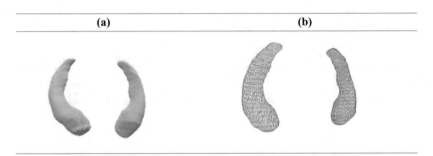

Fig. 5. Rendering of the right and left hippocampus after 3D reconstruction, (a) surface, (b) wireframe

After a 3D reconstruction phase, the volumes were calculated. The asymmetry between the right and left hippocampus was quantified and an asymmetry index (AI, Eq. 1) was estimated.

As mentioned earlier, lateralization of the epileptogenic zone is a crucial step in diagnosing epilepsy and improving therapy. Indeed, the accuracy of the surgical procedure depends on the accuracy of the localization of the epileptogenic zone.

Segmented images were realigned in three dimensions and reconstructed into entire contiguous coronal slices of 2 mm thickness.

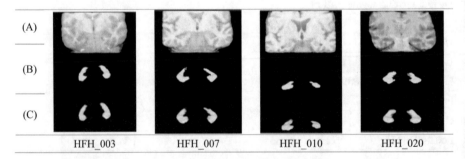

Fig. 6. Example of 3D reconstruction results (A) Original image (B) 3D reconstruction of manual segmentation, (C) 3D reconstruction of DL segmentation.

To better interpret the results, we performed a comparative study between the hippocampal volumes reconstructed from the ground truth images and those reconstructed from the automatically segmented images Fig. 6 and Table 4.

Table 4. Volume comparison, manual segmentation and automatic segmentation, R H V: Right hippocampus volume, L H V: Left Hippocampus volume

Subject number	Ground truth			U-Net segmentation		
	R H V (mm^3)	L H V (mm^3)	AI	R H V (mm^3)	L H V (mm^3)	AI
HFH_003	1427.05	2217.57	43.38%	1416.53	2326.25	48.61%
HFH_007	2245.85	2038.16	9.7%	2318.39	1976.13	15.94%
HFH_010	2272.77	859.50	90.24%	2361.74	907.19	88.99%
HFH_020	2786.54	2550.09	8.86%	2752.92	2601.42	5.66%

4 Discussion

The study introduced a comprehensive process to automatically segment the right and left hippocampi. The existing measures for expert radiological assessment of MRIs to detect mTLE related to hippocampal atrophy are inadequate as they are not easily evaluated. Therefore, we developed and explored the implementation of a convolutional neural network trained on preoperative MRIs to aid in diagnosing these conditions. Accurate normative volumetric information is necessary to identify structural volume reduction and alterations in left and right asymmetry in patients. The presence of asymmetric volumes in structures where symmetry is expected may indicate a pathological process [18].

This study confirms the results in the literature for mTLE patients using the hippocampus as the target volume [3, 4, 8,] showing that a pattern suggestive of mTLE can be demonstrated by MR images and finding the possibility of detecting unilateral hippocampal atrophy.

Our results showed that the dice coefficient was significantly increased using DL compared to other automated methods, and that segmentation accuracy was higher using our approach, leading to a higher concordance with the manual segmentation method in terms of volume calculation for the lateralisation of the epileptogenic side. Moreover, segmentation accuracy was higher with our approach, resulting in a higher agreement with the manual segmentation method in terms of volume calculation for the lateralization of the epileptogenic side. Lateralization accuracy achieved with our DL method was closer to the manual method compared to other methods.

In Table 4, we compared the volumes computed from the ground truth images with those produced by our approach. The results demonstrated close comparability. Our fast segmentation method requires only a fraction of the effort needed for manual segmentation. Comparing atrophy rates based on fully manual segmentation to our method showed promising possibilities for implementation in clinical workflow.

5 Conclusion

The Scheltens visual scale is employed in clinical settings to assess mesial temporal lobe atrophy. Quantitative volumetry techniques support the visual assessment of hippocampal volume. Our model could prove helpful in diagnosing preoperative hippocampal atrophy in mTLE patients.

This was achieved through the utilisation of 3D reconstruction, hippocampal volume computing and a focus on the hippocampus. This approach was shown to save time with the automatic MR imaging, while providing an accurate rendering of the anatomical and structural state of the hippocampus. Our evaluation metrics yielded results that surpassed those obtained in [11].

Patients with mesial temporal lobe epilepsy (mTLE) indicated the potential use of AI as a biomarker for this disease. Although volumetric analysis may be insufficient to determine laterality in mTLE, it remains an essential aspect of the investigation process. Therefore, automatic segmentation methods require continuous modification, particularly when there is a lack of obvious volumetric disparity between the hippocampi.

As part of our ongoing work, we are currently developing a new segmentation method to enhance performance. We aim to extend this approach to other structures that may be affected by atrophy in TLE, including the amygdala, parahippocampal gyrus, and its subdivisions. Furthermore, this study could be expanded to include other complementary techniques, such as simultaneous EEG/PET and/or EEG/FMRI acquisitions, to identify the electrical discharges and any associated hemodynamic or metabolic responses.

References

1. Jackson, G.D., Berkovic, S.F., Tress, B.M., Kalnins, R.M., Fabinyi, G.C., Bladin, P.F.: Hippocampal sclerosis can be reliably detected by magnetic resonance imaging. Neurology **40**, 1869 (1990)
2. Jack, C.R., Jr.: MRI-based hippocampal volume measurements in epilepsy. Epilepsia **35**, S21–S29 (1994)
3. Huijin, H., Tianzhen, S., Xingrong, C.: Comparison of MRI, MRS, PET and EEG in the diagnosis of temporal lobe epilepsy. Chin. Med. J. **114**, 70–9 (2001)
4. Suzuki, M., et al.: Male-specific volume expansion of the human hippocampus during adolescence. Cereb. Cortex **15**, 187–193 (2005)
5. Babb, T.L.: Pathological findings in epilepsy. Surgical treatment of the epilepsies, pp. 511–540 (1987)
6. Folle, L., Vesal, S., Ravikumar, N., Maier, A.: Dilated Deeply Supervised Networks for Hippocampus Segmentation in MRI. In: Handels, H., Deserno, T., Maier, A., Maier-Hein, K., Palm, C., Tolxdorff, T. (eds.) Bildverarbeitung für die Medizin 2019. Informatik aktuell. Springer Vieweg, Wiesbaden (2019). https://doi.org/10.1007/978-3-658-25326-4_18
7. Liu, M., et al.: A multi-model deep convolutional neural network for automatic hippocampus segmentation and classification in Alzheimer's disease. Neuroimage **208**, 116459 (2020)
8. Yao, W., Wang, S., Fu, H.: Hippocampus segmentation in MRI using side U-net model. In: Conference 2019, on Neural Information Processing (2019)
9. Pardoe, H.R., Pell, G.S., Abbott, D.F., Jackson, G.D.: Hippocampal volume assessment in temporal lobe epilepsy: how good is automated segmentation? Epilepsia **50**, 2586–2592 (2009)

10. Hammers, A., et al.: Automatic detection and quantification of hippocampal atrophy on MRI in temporal lobe epilepsy: a proof-of-principle study. Neuroimage **36**, 38–47 (2007)
11. Akhondi-Asl, A., Jafari-Khouzani, K., Elisevich, K., Soltanian-Zadeh, H.: Hippocampal volumetry for lateralization of temporal lobe epilepsy: automated versus manual methods. Neuroimage **54**, S218–S226 (2011)
12. Ronneberger, O., Fischer, P., Brox, T.: U-Net: convolutional networks for biomedical image segmentation. In: International Conference on Medical Image Computing and Computer-Assisted Intervention; Munich, Germany, pp. 234–241 (2015)
13. Guan, X., et al.: 3D AGSE-VNet: an automatic brain tumor MRI data segmentation framework. BMC Med. Imaging **22**(1), 1–18 (2022). https://doi.org/10.1186/s12880-021-00728-8
14. Lorensen, W.E., Cline, H. E.: Marching cubes: a high resolution 3D surface construction algorithm. ACM SIGGRAPH Comput. Graph. **21**, 163–169 (1987)
15. Rohmer, D.: Les surfaces gagnent du volume. Quadrature (2014)
16. Pedraza, O., Bowers, D., Gilmore, R.: Asymmetry of the hippocampus and amygdala in MRI volumetric measurements of normal adults. J. Int. Neuropsychol. Soc. **10**, 664–678 (2004)
17. Fischl, B., et al.: Whole brain segmentation: automated labeling of neuroanatomical structures in the human brain. Neuron **33**, 341–355 (2002)
18. Watson, C., Jack, C.R., Jr., Cendes, F.: Volumetric MRI: clinical applications and contributions to the understanding of TLE. Arch. Neurol. **54**, 1521–1531 (1997)

Integrating Explainable AI: Breakthroughs in Medical Diagnosis and Surgery

Ana Henriques[1], Henrique Parola[1], Raquel Gonçalves[2], and Manuel Rodrigues[1]([⊠])

[1] LASI/ALGORITMI Centre, University of Minho, Guimarães, Portugal
{pg50196,pg50415}@alunos.uminho.pt,
manuel.rodrigues@algoritmi.uminho.pt
[2] Hospital da Senhora da Oliveira Guimarães, EPE Portugal, Guimarães, Portugal

Abstract. The challenge of explaining the results generated by artificial intelligence (AI) is a significant obstacle to their widespread acceptance, which is why increased attention has been paid to the explainability in AI (XAI) in recent years. Given its impact on the medical sector, the survey seeks to demonstrate the important role of XAI in ensuring the reliability and accountability of AI in the domains of diagnosis and surgery. Therefore, we conduct an in-depth look at the applications and challenges of XAI in these areas by reviewing articles published between 2022 and 2023. The survey aims to explore the categorization of XAI techniques, establish their taxonomy, address trade-offs between model performance and interpretability and emphasize the importance of achieving a balance in practical applications. The findings of this study confirm the potential of XAI in medicine as a promising avenue for exploration, providing guidance for the development of medical XAI applications.

Keywords: Artificial Intelligence · Explainable Artificial Intelligence · XAI · Machine Learning · Deep Learning · Medicine · Diagnosis · Surgery

1 Introduction

Artificial intelligence (AI) is no longer just a nascent technology, but a transformative force that is profoundly changing the global paradigm. The intersection of AI and medicine is particularly relevant, with the main goal of medical AI research being the development of intelligent systems, such as machine learning (ML) algorithms, to support physicians in clinical decision-making by automating tasks [1]. However, concerns arise from the opacity of millions of parameters in deep learning (DL) models, as they provide final decisions without transparent explanations, notably because of their black-box nature. Jacovi *et al.* [2] emphasized the contractual nature of the agreement between humans and AI systems, requiring explicit interaction to foster trust in their relationship. The lack of transparency compromises trust among healthcare professionals and acceptance in the medical community, demanding a shift in focus from the model accuracy to the explainability of AI in the clinical setting.

© The Author(s), under exclusive license to Springer Nature Switzerland AG 2024
Á. Rocha et al. (Eds.): WorldCIST 2024, LNNS 986, pp. 254–272, 2024.
https://doi.org/10.1007/978-3-031-60218-4_23

1.1 XAI-Related Concepts

According to Owens *et al.* [3], explainable AI (XAI) facilitates the transfer of understanding to the end user by underscoring important decision paths in the model. This enables interpretability by humans at different stages of the process while maintaining predictive accuracy. XAI techniques can be categorized into intrinsic and post-hoc methods. The former involve the adoption of inherently interpretable ML models to understand decision-making processes without additional information. However, these are limited to specific models and may have lower predictive performance. In contrast, post-hoc methods can be applied to any ML algorithm and generate explanations after training the model [4]. Furthermore, XAI methods can be either model-agnostic, by revealing the decision process of any black-box model, or model-specific, by being tailored to specific algorithm features. XAI techniques are further categorized based on interpretability, with global explanations capturing general relationships learned by neural networks and local explanations providing insights into individual inputs [5]. Well-known XAI post-hoc methods include SHapley additive exPlanations (SHAP), local interpretable model-agnostic explanations (LIME) and gradient-weighted class activation mapping (Grad-CAM).

1.2 Contributions to the Medical Field

Concerns arise among healthcare experts regarding the black-box nature of current DL models for medical image analysis. Our survey aimed to offer both healthcare professionals and AI experts with valuable insights into the latest advances in medical XAI methods in clinical settings concerning the domains of diagnosis and surgery. Therefore, four key research questions (RQs) will be strategically addressed: (RQ1) What are the current trends in XAI techniques in the medical domain?; (RQ2) How are studies navigating the delicate trade-off between accuracy and explainability in medical AI applications?; (RQ3) Which AI models are encompassed by which XAI methods?; and (RQ4) What future directions and recommendations are emerging in the field of diagnosis and surgery at the intersection of AI and XAI? Extending the work begun by Zhang *et al.* [6], our survey not only consolidates the current state of knowledge on this topic, but also charts a roadmap for future research to promote advances that uphold the dual imperatives of accuracy and explainability in medical AI systems.

2 Applications of XAI in Medicine

2.1 Methods

The selection of articles for the survey was conducted through a literature review adapted from the Preferred Reporting Items for Systematic Reviews and Meta-Analyses (PRISMA) guidelines, considering the parameters from the prior investigation in the study area [6], introducing additional considerations to ensure a more comprehensive collection of information.

Search Strategy. Only conference papers, journal articles or clinical trials published between 2022 and 2023, in English and from the following databases were considered for the survey: PubMed, IEEE Xplore, Web of Science, Science Direct and Scopus. The search string used was: ("All Metadata": "Explainability" OR "All Metadata": "Explainable" OR "All Metadata": "XAI") AND ("All Metadata": "AI" OR "All Metadata": "Artificial Intelligence") AND ("All Metadata": "Deep Learning" OR "All Metadata": "Machine Learning") AND ("All Metadata": "Diagnosis" OR "All Metadata": "Surgery").

Selection Criteria. After an initial screening of the studies eligible for the survey, the results underwent a selection process based on their title and abstract. The inclusion and exclusion criteria of the selected articles were: (i) dealing with at least one AI model, (ii) including at least one XAI technique, (iii) focusing on diagnosis or surgery, and (iv) proposing a solution to a classification problem.

Data Extraction. The literature review was conducted to extract the following data: (i) the research aim, (ii) the AI algorithm, (iii) the performance metrics of the algorithm, (iv) the XAI techniques used, and (v) the XAI scoring metrics.

2.2 Results

Table 1 summarizes the data extracted from the reviewed literature on using XAI techniques in the medical field, with 89.3% concentrating their efforts on diagnosis [7–31] and 10.7% on surgery [32–34]. Interestingly, two of the diagnosis-oriented investigations [27, 32] briefly acknowledged the potential utility of XAI in the surgical context. The AI evaluation metrics presented in Table 1 are linked to the optimal model.

The predominant medical objective was covered by 28.6% of the studies, revolving around research into XAI in relation to cancer [11, 14, 21–23, 29, 32, 34]. Within this overarching theme, breast cancer received significant attention (10.7%) [14, 29, 34], followed by articles looking at the intricacies of lung cancer (7.1%) [11, 21]. Further analysis revealed that other prevalent targets, each accounting for 10.7%, included COVID-19 [13, 21, 30], a contemporary healthcare concern, and medical imaging modalities [7, 20], such as X-rays [7, 20] and endoscopy [28]. In a nuanced distribution, 7.1% of the studies focused on either Parkinson's disease [17, 18], pneumonia [13, 19], attention deficit hyperactivity disorder (ADHD) [12, 15], sleep apnea [9, 31], skin lesions [10, 26] or cardiovascular diseases [16, 24]. Individual studies have addressed other issues ranging from hip prosthetic failure [27] and glaucoma [33] to diabetes [8] and chronic kidney disease (CKD) [25]. These percentages are not mutually exclusive, as some studies pursued more than one objective, AI algorithm and XAI method.

In the realm of AI models, algorithms, and frameworks, 64.3% of the studies incorporated different architectures of convolutional neural networks (CNNs) [7, 10–13, 15, 17–21, 24, 26–28, 30, 31, 34]. ResNet was the frontrunner with 21.4% [10, 20, 26, 28, 30, 34], while VGG [17, 20, 28] and Inception [17, 20, 28] each commanded a 10.7% share. In addition, 21.4% embraced the combination of a CNN-based model with transfer learning [10, 19, 20, 26–28]. Securing the runner-up position, 39.3% of the studies employed random forests (RFs) [8, 9, 22–25, 29, 32, 33]. In parallel, 32.1% used gradient boosting machine (GBM) [9, 16, 23–25, 29, 33, 34], with extreme gradient boosting

Table 1. Literature review of medical XAI applications in diagnosis and surgery

Study	Diagnosis or Surgery?	Aim	AI Algorithm	AI Evaluation Metrics	XAI Technique	XAI Evaluation?
Pham *et al.* (2022) [7]	Diagnosis	Chest radiographs scans	CNN-based models: EfficientNet-B6 and EfficientDet-B6	Specificity: 90.0% Sensitivity: 93.3% F1-score: 63.1% AUC: 96.7%	Saliency Maps	No
Vishwarupe *et al.* (2022) [8]	Diagnosis	Diabetes	Random Forest (RF)	Accuracy: 82.23%	SHAP, LIME	No
García *et al.* (2022) [9]	Diagnosis	Sleep apnea	RF, Logistic regression, Decision tree (DT), K-nearest neighbors (KNN), Gradient boosting classifier (GBM)	Accuracy: 70.7% ROC-AUC: 72.8% F1-score: 39.3%	LIME	No
Nigar *et al.* (2022) [10]	Diagnosis	Skin lesion	CNN-based model: ResNet18 with Transfer Learning	Accuracy: 94.47% Precision: 93.57% Recall: 94.01% F1-score: 94.45%	LIME	No
Masot *et al.* (2022) [11]	Diagnosis	Lung cancer	3-class Color CNN, 3-class Grey-scale CNN, and 2-class CNN	BNG (Color—Grey) Acc: 99.6%—97.5% Prec: 99.7%—96.3% Sensit: 99.7%—96.2% F1: 99.4%—96.2%	Grad-CAM, Occlusion sensitivity	No

(*continued*)

Table 1. (*continued*)

Study	Diagnosis or Surgery?	Aim	AI Algorithm	AI Evaluation Metrics	XAI Technique	XAI Evaluation?
Cozma et al. (2022) [32]	Surgery	Thoracic cancer	RF, Support vector machine (SVM), GAN	N/A	SHAP, LIME	No
Loh et al. (2023) [12]	Diagnosis	ADHD and Conduct Disorder (CD)	1D CNN classifier: 10-fold CNN	Accuracy: 96.04% Precision: 96.26% Sensitivity: 95.99% F1-score: 96.11%	Grad-CAM	No
Zou et al. (2023) [13]	Diagnosis	Pneumonia and COVID-19	CNN: Xception	AUC: 89,9% Accuracy: 89,9%	SHAP, Grad-CAM, Grad-CAM + +, LIME, Saliency Map	Yes
Khater et al. (2023) [14]	Diagnosis	Breast cancer	Artificial neural network (ANN), KNN	KNN—ANN Acc: 97.7%—98.6% Prec: 98.2%—94.4%	PFI, PDP, SHAP	No
Yilmaz et al. (2023) [16]	Diagnosis	Acute myocardial infarction (AMI)	RF, XGBoost, AdaBoost, Light Gradient Boosting Machine (LGBM)	Accuracy: 83% Precision: 85% F1-score: 82% AUC: 92%	SHAP	No
Saravanan et al. (2023) [17]	Diagnosis	Parkinson's disease (PD)	CNN-based models with transfer learning: VGG19 Net and Inception	Accuracy: 98.45%	LIME	No
Camacho et al. (2023) [18]	Diagnosis	PD	Simple fully CNN, Nu-SVM with a radial basis kernel function	Accuracy: 79.3% Precision: 80.2% Specificity: 81.3% Sensitivity: 77.7% AUC-ROC: 0.87	SmoothGrad	No

(continued)

Table 1. (*continued*)

Study	Diagnosis or Surgery?	Aim	AI Algorithm	AI Evaluation Metrics	XAI Technique	XAI Evaluation?
Lysdahlgaard et al. (2023) [20]	Diagnosis	Wrist and elbow radiographs	CNN + transfer learning: VGG16, VGG19, ResNet50V2, ResNet101V2, ResNet152V2, DenseNet121, DenseNet169, DenseNet201, InceptionV3, and Xception	Accuracy (wrist radiographs): 81% Accuracy (elbow radiographs): 60%	Grad-CAM	No
Islam et al. (2023) [21]	Diagnosis	Lung cancer and COVID-19	CNN- and RNN-based models: DCNN and Gated recurrent unit (GRU)	Cancer—COVID-19 Acc: 98.79%—99.3% Prec & Recall & F1: 99% AUC: 0.99—0.99 Specificity: 99%—99.5%	Grad-CAM, LIME, SHAP	No
Hossain et al. (2023) [24]	Diagnosis	Cardiovascular disease	SVM, DT, KNN, Naïve Bayes (NB), Multilayer Perceptron (MLP), CatBoost, GBM, AdaBoost, RF and hybrid CNN with LSTM network	CNN-LSTM Accuracy: 74.15% Sensitivity: 72.04% Specificity: 77.11% Precision: 81.82% Recall: 72.04% F-measure: 76.62%	SHAP	No
Bellantuono et al. (2023) [23]	Diagnosis	Thyroid cancer	RF, XGBoost, SVM, Gaussian Naïve Bayes	AUC: 94.41%, 92.71%, 92.12% and 93.12%, respectively	SHAP	No

(*continued*)

Table 1. (*continued*)

Study	Diagnosis or Surgery?	Aim	AI Algorithm	AI Evaluation Metrics	XAI Technique	XAI Evaluation?
Sánchez et al. (2023) [25]	Diagnosis	Chronic kidney disease	RF, Extra Trees, AdaBoost and XGBoost	DT&AdaBoost Accuracy: 98.3% Sensitivity: 97.3% Specificity: 100% F1-score: 98.6% Precision: 100%	SHAP, PFI, PDP	Yes
Caballero et al (2023) [15]	Diagnosis	ADHD	CNN	N/A	Occlusion maps	No
To et al (2023) [34]	Surgery	Breast cancer	XGBoost, CNN-based model: ResNet50 and DenseNet169	Accuracy: 95.0% Sensitivity: 100% Specify: 87.3%	Grad-CAM + +	No
Muscato et al (2023) [27]	Diagnosis	Hip prosthetic failure	SVM, CNN-based model with transfer learning: Densenet169	Specificity: 86.3% Recall: 91.9% Precision 95.6% AUC: 96.1% F1-score: 87.4% Accuracy: 86.1%	SHAP	No
Mena et al (2023) [22]	Diagnosis	Prostate cancer	RF, KNN, rpart	RF—KNN—rpart Sens: 90%—92%—85% Spec: 92%—99%—97% AUC: 96%—94%—87% F1: 76%—61%—59%	SHAP	No

(continued)

Table 1. (*continued*)

Study	Diagnosis or Surgery?	Aim	AI Algorithm	AI Evaluation Metrics	XAI Technique	XAI Evaluation?
Nayak et al (2023) [26]	Diagnosis	Monkeypox	CNN models with transfer learning: GoogLeNet, Places365-GoogLeNet, SqueezeNet, AlexNet and ResNet-18	Accuracy: 99.49% Sensitivity: 99.43% Specificity: 100% Precision: 100% F1-score: 99.49%	LIME, Grad-CAM	No
Sheu et al (2023) [19]	Diagnosis	Pneumonia	Deep CNN with transfer learning	Accuracy: 93.29% AUC: 95.43%	SHAP, IF-THEN rules	No
Varam et al. (2023) [28]	Diagnosis	Endoscopy imaging	Transformer- and CNN models with transfer learning: Vision Transformer, MobileNetv3Large, ResNet152v2, InceptionV3, EfficientNet, VGG16 and VGG19	F1-score: 97 + -1%, 95 + -2%, 94 + -1%, 26 + -4%, 95 + -1%, 94 + -2% and 38 + -6%, respectively	Grad-CAM, Grad-CAM + +, LayerCAM, LIME, SHAP	No
Tao et al (2023) [33]	Surgery	Glaucoma	Cox proportional hazards (CPH), Random Survival Forest, Gradient-boosting survival, Deep feed-forward neural network: DeepSurv	DeepSurv C-index: 77.5% Mean AUC: 80.2%	SHAP	No

(*continued*)

Table 1. (*continued*)

Study	Diagnosis or Surgery?	Aim	AI Algorithm	AI Evaluation Metrics	XAI Technique	XAI Evaluation?
Massafra et al (2023) [29]	Diagnosis	Breast cancer	XGBoost, RF, SVM, Naïve Bayes	RF-XGB Cohen's kappa: 0.75 ± 0.17 and 0.7 ± 0.16 at 5-year and 10-year follow-ups, respectively	SHAP	Yes
Nkengue et al (2024) [30]	Diagnosis	COVID-19	CNN-based model ResNet and LSTM	Sensitivity: 96.48% Precision: 96.48% Specificity: 96.24% Accuracy: 96.48%	Grad-CAM + +	No
García et al (2024) [31]	Diagnosis	Sleep apnea	CNN, RNN, Linear regression	Accuracy: 93.5%—90.5% Sensit: 71.4%—78.3% Specif: 97%—93.8% ICC = 0.9465—0.9004	Grad-CAM	No

(XGBoost) leading with 20.7% [16, 23, 25, 29, 34]. Support vector machine (SVM) found application in 24.1% of the studies [18, 23, 24, 27, 29, 32], whereas decision trees (DTs) [9, 22, 24, 25] and k-nearest neighbors (KNNs) [9, 14, 22, 24] were tied at 13%. Within this diverse landscape, Naïve Bayes was implemented in 10.7% of the studies [23, 24, 29], while Long Short-Term Memory (LSTM) [24, 30] and Recurrent Neural Networks (RNN) [21, 31] together comprised 7.1%.

A remarkable 53.6% of the studies demonstrated a preference for SHAP [8, 13, 14, 16, 19, 21–25, 27–29, 32, 33], while 7.1% each employed permutation feature importance (PFI) [14, 25] and PDP [14, 25]. Considering that PFI and PDP were consistently utilized in conjunction with SHAP, a cumulative 53.6% of research contributed to the domain of feature relevance explanation. In contrast, 39.3% of the studies opted for Grad-CAM [11, 13, 20, 21, 26, 28, 31, 32], incorporating various extensions such as Grad-CAM + + [13, 28, 30, 34], SmoothGrad [18] and LayerCAM [28], and an additional 7.1% harnessed saliency maps [7, 13]. In total, 46.4% took on the task of visual explanation. As the third most predominant technique, 32.1% selected LIME [8–10, 13, 17, 21, 26, 28, 32], which indicates that explanation by simplification was addressed in the same proportion. Furthermore, single studies integrated IF-THEN rules [19] and occlusion sensitivity [11], thereby introducing rule- and perturbance-based explanations, respectively. Only 14.3% of the articles included specific strategies to evaluate the effectiveness of XAI [13, 20, 25, 29].

2.2.1 Diagnosis

Grad-CAM. Lysdahlgaard *et al.* [20] investigated different transfer learning models using Grad-CAM on wrist and elbow radiographs. The study aimed to assist clinicians in detecting limb abnormalities by using well-known CNN-based models such as VGG, ResNet, DenseNet, Inception and Xception. GradCAM modifies the final convolutional layer in these models to recognize the importance of individual neurons, generating heat maps that visually explain the model's predictions. Nayak *et al.* [26] also applied transfer learning using the same XAI technique in various CNN-based models: ResNet, GoogLeNet, SqueezeNet, AlexNet. Their goal was to diagnose monkeypox from skin lesion images. Grad-CAM explanations indicated warmer regions as more influential in predictions, whereas cooler areas are less significant.

The research by Loh *et al.* [12] classified electrocardiogram (ECG) segments with 1-dimensional CNN models to detect ADHD, CD and ADHD + CD. A particular focus was placed on the application of Grad-CAM to ECG data, which yielded temporal localization of critical features that support the diagnosis of these conditions. The method effectively improved the classification accuracy from 87.19% to 96.04% and provided better model interpretability. Masot *et al.* [11] implemented three CNN classifiers to diagnose lung cancer from color and grayscale images. They addressed the black-box problem of DL with the employment of Grad-CAM and occlusion sensitivity, where the second mechanism assessed the robustness of the classifier by eliminating pixel areas and evaluating the reduction in accuracy. The final report for pathologists included a heat map superimposed on the original image, showing critical classification regions along with the result and a percentage of reliability.

Various versions of Grad-CAM, such as Grad-CAM++ and LayerCAM, have appeared in research. Grad-CAM++ addresses the challenge of Grad-CAM to accurately recognize objects among multiple instances of the same class [13] by incorporating second-order gradient computations [28]. On the other hand, LayerCAM generates class activation maps by using multiple convolutional layers, unlike Grad-CAM which relies only on the last convolutional layer [28]. For example, Nkengue et al. [30] focused exclusively on the classification of ECG signals using X-RCRNet, based on ResNet18, to diagnose COVID-19. In terms of model explainability, the study employed Grad-CAM++ to highlight features that influenced COVID-19 classification, showing that the ST interval[1] emerged as the feature with the highest gradient value.

Varam et al. [28] used Vision Transformer together with CNN-based transfer learning for the classification of endoscopy images. Due to the sensitivity of endoscopic imaging, global explanations for the model's predictions were insufficient. Customized diagnosis for specific cases is critical, requiring Grad-CAM for personalized clarity. The interest points made by Varam et al. [28] was the utilization Grad-CAM, Grad-CAM++ and LayerCAM to adapt a model-specific technique to Transformer models without convolutional layers. By focusing on the last token in the last attention layer, they effectively computed gradients even though there were no traditional convolutional layers. Among the tested methods, Grad-CAM with first-order gradients outperformed LIME and SHAP.

Similar to Lysdahlgaard et al. [20], Zou et al. [13] implemented Xception in combination with a fully connected network and innovated by providing an ensemble of different methods, including Grad-CAM++ and SHAP. Their study investigated whether the combination of these XAI methods produce complementary effects when applying kernel ridge regression to map discriminative regions, visually explaining a DL model for predicting mortality risk in patients with pneumonia and COVID-19. One distinguishing factor of the paper is the rigorous framework for evaluating XAI techniques through quantitative and qualitative metrics, including Decision Impact Ratio, Confidence Impact Ratio, Accordance recall, Accordance precision, Intersection over Union, and radiologists' trust. Quantitatively, the XAI ensemble outperformed other methods and garnered the highest trust among radiologists, receiving an average vote of 70.2%. The aforementioned [20] also developed an XAI evaluation method, using Dice similarity coefficient to quantify the similarity between the generated heat maps and a standard reference dataset with metal detection. This provided a robust measure to assess the agreement between the predicted and actual results, suggesting that the algorithms produce more consistent heat maps when interpreting conspicuous features like metal rather than fractures. Camacho et al. [18] implemented a modified version of the simple fully CNN to diagnose PD using T1-weighted magnetic resonance imaging datasets. Their DL model, $CNN_{Jacobians}$, demonstrated superior performance, being further analyzed with the SmoothGrad algorithm for saliency map calculation to gain a better understanding of the model's general behavior. The identified regions included subcortical areas such as the amygdala, putamen, and hippocampus as well as cortical regions such as the frontal and temporal cortex.

---

[1] The initial, slow phase of ventricular repolarization.

Although Grad-CAM visually identifies critical image regions that influence CNN predictions [21], other studies have extended Grad-CAM to architectures that combine CNN and additional modules. For instance, García *et al.* [31] integrated CNN and RNN to diagnose obstructive sleep apnea in children. With Grad-CAM heatmaps, their model focused on abrupt atrial fibrillation cessations and SpO2 drops to detect specific apneas and hypopnea desaturations, and often discarded patterns of hypopneas associated with arousals. Additionally, Islam *et al.* [21] combined the spatial feature extraction of CNN with the temporal dependency detection of GRU to improve the diagnosis of lung abnormalities and COVID-19. Using Grad-CAM, the model identified key disease indicators in CT scans for early detection of lung cancer, involving nodules, masses, and tumor patterns.

SHAP. While Grad-CAM is closely related to CNNs, some articles have explored traditional ML with the interpretation of SHAP due to its model-agnostic nature. Vishwarupe *et al.* [8] employed RF to predict diabetes in patients based on a clinical trial dataset, achieving 82% of accuracy using parameters such as age, skin thickness, body insulin, blood test and body mass index (BMI). Among the global SHAP interpretations, glucose level topped the list, followed by age, BMI, diabetes pedigree function, blood pressure, skin thickness and body insulin in determining diabetes. The model confirmed that increased insulin resistance drives up blood glucose in potential diabetes cases, which is consistent with clinical findings. Blood pressure, skin thickness and BMI had distinct effects due to the random nature of the studies, increasing the complexity of the dataset.

Similar to Vishwarupe *et al.* [8], Sánchez *et al.* [25] used RFs, but also Extra-Trees, AdaBoost and XGBoost with DT, to build an interpretable CKD prediction model. Ensemble trees are popular ML classifiers due to their robustness across different dataset sizes and good prediction performance. SHAP, in concordance with PDP and PFI, demonstrated that higher hemoglobin levels are related to lower CKD probability, and vice versa. High urine specific gravity levels also reduce the probability of CKD, whereas the presence of hypertension increases it. This study stood out in the literature by prioritizing a balance between performance and explainability. Through interpretability and fidelity metrics, they introduced the Fidelity-Interpretability index (FII) for comparative analysis of model explainability by relating the number of selected features and accuracy performance. XGBoost had the highest interpretability (88%) with three features (hemo, htn, sg), but every model achieved fidelity close to 100%.

Yilmaz *et al.* [16] employed SHAP in their experiment, which used RFs to assess the importance of hematological blood parameters in the prediction of acute myocardial infarction (AMI), developing an accurate and efficient diagnostic model. The XGBoost classifier was also selected based on its ability to handle complex datasets, while the AdaBoost classifier combined DTs to create a robust ensemble model. The LGBM was utilized to develop a classifier for the early detection of AMI. The SHAP chart-based analysis of LGBM reveals crucial hematological parameters for AMI and STEMI diagnosis, providing specific risk factors that go beyond descriptive statistical methods.

In addition to interpreting the RF model, Mena *et al.* [22] applied SHAP to gain information into the predictions of KNN and rpart. The study intended to unravel the complex behavior of these models using post-hoc techniques to calculate the influence of each gene on the individual predictions. The researchers found that increased expression of certain genes, such as PTEN, was associated with an increased risk of prostate cancer, while others, such as FOXA1, exhibited the opposite effect. Bellantuono *et al.* [23] investigated the application of SMOTE for each ML model, more precisely RF, XGBoost, SVM and Gaussian Naïve Bayes, to address variability in the diagnosis of thyroid cancer. Their performance was evaluated through the area under the curve (AUC) for a dataset with internal parameter optimization. SHAP clarified the impact of Raman spectral properties associated with the molecules on model predictions, highlighting oxidized cytochrome b, oxidized cytochrome c, carotenoids, among others.

Massafra *et al.* [29] adopted SHAP to offer detailed insights into feature relevance for breast cancer invasive disease events (IDE), over 5 and 10 years. They explored both global- and local-level explainability and distinguished between patients in the IDE and non-IDE classes. They assessed these predictions with RFs, SVMs, XGBoost and Naïve Bayes. Being XGBoost the best performing classifier, features such as Ki67, multiplicity and grading were portrayed as positive, while chemotherapy scheme and age were presented as evidence against the IDE prediction. SHAP was not only limited to explaining predictions for included patients, but was also employed for confounding patients, using the former as the training set and the latter as the test set.

Khater *et al.* [14] investigated KNN, SVM, in addition to XGBoost and tree-based methods. The best performing KNN model used SHAP to evaluate extreme features in breast cancer classification. The "area worst" feature, which represented the total area of the nucleus, was found to be the most influential and correlated with higher values with better prediction of malignant breast cancer. In another dataset, "bare nuclei" was the most important indicator of malignant breast cancer. Elevated levels, indicating the absence of cell membranes, strongly predicted malignancy. Researchers reflected that the "area worst" in one dataset could be influenced by abnormal cell growth or division, which can be seen in the "bare nuclei" in another dataset. This study introduced a systematic framework for explainable ML by incorporating a variety of XAI techniques. In contrast to previous studies that relied on a single XAI method, it emphasizes the advantages of combining SHAP, PDP and PFI.

Besides Grad-CAM, CNN models are frequently identified in the context of transfer learning using SHAP. Sheu *et al.* [19] developed an interpretable CNN-based model with transfer learning, DenseNet121, for the diagnosis of pneumonia. It utilized the VinDr open dataset to transfer knowledge, focusing on chest X-ray (CXR) features. This study stood out by integrating human-in-the-loop approach (feedback) into training and evaluation. Medical experts flag misclassified training images refining the model for improved accuracy. This feedback process ensured continuous improvement and adaptability to effectively treat disease and precisely identifies the critical medical features for accurate diagnosis. SHAP underlined the severity of pneumonia in the different CXR sections, supporting localization and treatment decisions. In this work, SHAP showed higher precision in localizing features compared to Grad-CAM.

The investigation by Varam *et al.* [28] also applied SHAP in a transfer learning paradigm. The authors emphasized the ability to attribute relevance to each feature and to consider the interactions between features as the reason for the algorithm's popularity. SHAP provided clear explanations for many classes in endoscopic images, but struggled with more complex inputs, leading to reduced image quality and difficulty in recognizing important regions to support predictions. Grad-CAM and SHAP were implemented in conjunction to provide a stable interpretation, outperforming the individual techniques by assigning low weight to the special area outside the lungs, not helpful for decision making due to presence of text, catheters, or lines in the X-ray image. In addition, Islam *et al.* [21] also benefited from SHAP and Grad-CAM by visualizing vital features in their datasets for multi-class lung cancer and COVID-19 classifications.

Hossain *et al.* [24] utilized SHAP in conjunction with a hybrid CNN-LSTM model to predict cardiovascular disease. By calculating the SHAP values, the researchers discerned informative features that influenced the model, considering all potential feature subsets, both with and without feature engineering. The analysis identified systolic blood pressure and age as the most important for predicting cardiovascular disease. Lastly, the study by Cozma *et al.* [32] focused on improving the diagnosis of lung cancer using an algorithm that involved a series of pipeline processing steps. It started with the development of an ensemble classifier that achieved an accuracy of over 72% on the test set after employing the SMOTE method for balancing. The next algorithmic step was the application of model-agnostic XAI techniques, LIME and SHAP, to provide both local and global coverage, after investigating the clinical parameters and key indicators. Despite limitations in data availability, the study demonstrated a 7% improvement in the test set compared to medical decisions alone, showcasing the potential of integrating XAI in real-life diagnosis for thoracic surgery.

LIME. To demystify the VGG19-INC model used in the diagnostic accuracy of PD, Saravanan *et al.* [17] employed LIME. Firstly, transfer learning is preferred over starting from scratch with a limited dataset to reduce overfitting, enhance diagnostic accuracy and speed up convergence. However, it can be difficult to understand how these models arrive at their predictions. Their focus was to understand super pixels in spiral and wave drawings affecting PD prediction and spotting misclassification causes. In response, LIME created super pixel boundaries in the input, measuring the difference between the predicted and actual feature maps, and produced color-coded areas with features that influence on classification outcomes from the distance measured.

Existing algorithms require manual feature extraction, preprocessing and solely compute numerical values. To bypass these arduous stages and enable the algorithm to autonomously perform feature extraction, Nigar *et al.* [10] justified the use of transfer learning algorithm ResNet18 and an XAI based skin lesion classification system to improve the accuracy. Due to the intricacy of digital pictures of skin lesions and the high similarity in early stages of skin cancer, the detection of malignancy by visual evaluation became complicated, resulting in delayed diagnosis.

Varam *et al.* [28] compared LIME, Grad-CAM-based techniques and SHAP with different DL models for image recognition. In this article, SHAP faced the challenge of recognizing vital regions in endoscopic images, especially for complex inputs such as foreign body identification. LIME offered clearer explanations that SHAP could not achieve and provided more sensible explanations. However, utilizing a surrogate model with LIME may overlook essential features and accentuate black edges in endoscopic images, posing a limitation due to inaccurate generalizations. Nayak *et al.* [26] addressed this issue by incorporating CNN-based transfer learning models. In their approach, a regression tree served as the local approximation model to diagnose monkeypox from skin lesion images, presenting super pixels as LIME results.

García *et al.* [9] utilized several traditional ML models such as Logistic Regression, KNN, DT, RF and Gradient Boosting classifier. The data analyzed corresponded to EEG, ECG and respiration signals labeled with the existing episode of apnea and the patient's stage of sleep. Because of the signal-based nature of sleep apnea detection, the LIME graph lacked complete clarity. This reinforced the importance of identifying minimum and maximum points in detecting apnea, since it disrupts the normal breathing patterns of nasal airflow. Vishwarupe *et al.* [8] discovered that LIME identified glucose levels as a key factor in predicting diabetes, but this observation held true exclusively for individuals under the age of 45. In patients aged 25 years, factors such as BMI, blood pressure and age played a greater role in diagnosis, highlighting the need for different explanatory approaches such as the localized perspective of LIME for multiple insights. In addition to Grad-CAM, the results of LIME also enabled visual representations through heatmaps that showed the decision-making process. This was noticed by Nayak *et al.* [26] and by Islam *et al.* [21], where they focused on the five key features of LIME and used heatmaps to visually validate their ML model. These visualizations provided information into the image areas that impact predictions and revealed patterns within different COVID-19 and lung cancer classes.

Saliency Maps. Pham *et al.* [7] presented VinDr-CXR, a DL framework that utilized EfficientNet for classification of common lung diseases and abnormalities detection in CXR images. The interpretability of VinDr-CXR was demonstrated through saliency maps, which highlighted regions on CXRs that were relevant to abnormal patterns. The research findings pointed to potential applications in the automation of disease screening, supporting the generalizability of the model across geographical settings and its integration into various clinical workflows.

2.2.2 Surgery

SHAP. Diagnosis and surgery are intricately linked in certain instances, whether in determining the need for surgery or in post-operative analysis. Muscato *et al.* [27] employed radiographs obtained during routine follow-up after surgery to develop a DL approach for automatic hip prosthesis failure detection from conventional radiographs. Deep features were extracted via a pre-trained DenseNet169 CNN followed by SVM classification. As expected, the SHAP analysis highlighted the most important features mainly within the original image. Only two features were identified outside the original image, which reinforced the pivotal role of features extracted directly from the original images in defect detection.

Additionally, Cozma *et al.* [32] determined the optimal time for thoracic surgery by utilizing an XAI technique trained on a knowledge base. The resulting intelligent system combined the precision of expert systems with the flexibility of deep neural networks, seeking minimal human misinterpretations and an accurate estimate for surgical intervention timing. Tao *et al.* [33] developed AI survival models to predict the time of glaucoma progression to surgery, including regression, tree-based (RSF, GBS) and DL (Deep-Surv) approaches. The Cox proportional hazards model assessed risk using hazard rates, while the tree-based models incorporated ensemble techniques to enhance predictive accuracy and DeepSurv employed a deep feed-forward neural network for hazard estimation. The study examined the models' interpretability through SHAP and confirmed that critical predictors were consistent across models.

Grad-CAM. To address the problem of 'vanishing gradient' in CNNs during brain tumor surgery, To *et al.* [34] employed transfer learning by using the features of a pre-trained ResNet50 for patch classification via XGBoost. They tackled overfitting with Grad-CAM + + ap-plied to a pre-trained DenseNet169 model and explained model decisions by visualizing weighted feature maps. The method successfully localized regions of interest in deep ultraviolet (DUV) images, exhibiting 95% accuracy for DUV whole-surface im-aging (WSI) and 100% sensitivity for malignant cases. Compared to a ResNet50-based approach, the proposed ensemble method integrating XGBoost and Grad-CAM + + showed increased accuracy in the assessment of breast cancer margins.

2.3 Discussion

Between 2019 and 2023, CNNs and RFs emerged as the prominent AI models when integrating XAI techniques. Within CNNs, there has been a notable 28% increase in usage from 2022 to 2023 when compared to the preceding period from 2019 to 2021. Although VGG-16 was recognized by Zhang *et al.* [6] as the most used CNN-based model in transfer learning paradigms, our research reveals a growing prevalence of ResNet. Current trends in XAI highlight SHAP, Grad-CAM and LIME as primary techniques, diverging from the prior dominance of LIME between 2019 and 2021. SHAP has now surpassed LIME in popularity, featuring in over 20% of recent studies. The landscape of Grad-CAM-based methods shows diversity in implementations, encompassing Grad-CAM + +, LayerCAM and SmoothGrad. Despite these advancements, there is a lack of articles evaluating XAI methods, both quantitatively and qualitatively, over the last four years. Various metrics, such as Decision Impact Ratio, Confidence Impact Ratio and DSC, exist but are still relatively unexplored in this evolving area.

A wide range of medical research purposes was revealed over the past four years. Cancer, particularly breast and lung cancer, emerged as the most frequently discussed topic, although without significant relevance. In the period spanning 2019 to 2021, Alzheimer's disease claimed the second most discussed position, while in the past two years, COVID-19 has become a common theme in the literature, coexisting with discussions on cancer. AI applications in healthcare face a significant challenge concerning interpretability. It remains imperative for a model to explain the rationale behind its predictions or recommendations, even when demonstrating high accuracy, as emphasized by Saravanan *et al.* [17]. Despite the utilization of XAI techniques, there is a lack of studies addressing the intricate trade-off between explainability and performance. In addition, Sánchez *et al.* [25] stood out by evaluating their best model not based on the optimal performance, but on the balance between explainability and accuracy, providing their own metric to estimate this balance.

Regarding future directions, the survey indicates a growing emphasis on the robust construction of XAI technique frameworks. This evolution is evident in several key methodologies by adding (i) both quantitative and qualitative metrics to assess the techniques' efficacy [13], (ii) ensemble of XAI techniques [14, 28] and (iii) the integration of feedback-based systems, adopting a human-in-the-loop approach [19]. These strategies not only contribute to the development of more robust and trustworthy models, but also assist in ensuring that the system is up to date with training updates.

3 Conclusion

In summary, our survey, conducted between 2022 and 2023, provided a comprehensive overview of the applications of XAI in the medical domain, particularly in diagnosis and surgery. We carefully examined each selected study, delving into the intricacies of the AI algorithms used, the XAI techniques and the types of explanations involved. Despite remarkable progress in this rapidly evolving field, there is still a significant gap in XAI assessments over the past four years. Recognizing the urgent need for transparent models that bridge the gap between AI and medical professionals, our survey foresaw an increasing focus on the construction of robust XAI frameworks. Our study aimed to contribute valuable insights to the area under investigation, guiding future research efforts to develop effective and transparent medical AI applications.

Acknowledgement. This work has been supported by FCT – Fundação para a Ciência e Tecnologia within the R&D Units Project Scope: UIDB/00319/2020.

References

1. Alloghani, M., Al-Jumeily, D., Aljaaf, A.J., Khalaf, M., Mustafina, J., Tan, S.Y.: The application of artificial intelligence technology in healthcare: a systematic review. In: Khalaf, M.I., Al-Jumeily, D., Lisitsa, A. (eds.) ACRIT 2019. CCIS, vol. 1174, pp. 248–261. Springer, Cham (2020). https://doi.org/10.1007/978-3-030-38752-5_20
2. Jacovi, A., Marasović, A., Miller, T., Goldberg, Y.: Formalizing trust in artificial intelligence: prerequisites, causes and goals of human trust in AI. In: FAccT 2021: Proceedings of the 2021 ACM Conference on Fairness, Accountability, and Transparency (2020)
3. Owens, E., Sheehan, B., Mullins, M., Cunneen, M., Ressel, J., Castignani, G.: Explainable artificial intelligence (XAI) in insurance. Risks **10**(12), 230 (2022). https://doi.org/10.3390/risks10120230
4. Zhang, C.A., Cho, S., Vasarhelyi, M.: Explainable artificial intelligence (XAI) in auditing. Int. J. Acc. Inf. Syst. **46**, 100572 (2022). https://doi.org/10.1016/j.accinf.2022.100572
5. van der Velden, B.H.M., Kuijf, H.J., Gilhuijs, K.G.A., Viergever, M.A.: Explainable artificial intelligence (XAI) in deep learning-based medical image analysis. Med. Image Anal. **79**(102470), 102470 (2022). https://doi.org/10.1016/j.media.2022.102470
6. Zhang, Y., Weng, Y., Lund, J.: Applications of explainable artificial intelligence in diagnosis and surgery. Diagnostics (Basel) **12**(2), 237 (2022). https://doi.org/10.3390/diagnostics12020237
7. Pham, H.H., Nguyen, H.Q., Nguyen, H.T., Le, L.T., Khanh, L.: An accurate and explainable deep learning system improves interobserver agreement in the interpretation of chest radiograph. IEEE Access **10**, 104512–104531 (2022). https://doi.org/10.1109/ACCESS.2022.3210468
8. Vishwarupe, V., Joshi, P.M., Mathias, N., Maheshwari, S., Mhaisalkar, S., Pawar, V.: Explainable AI and interpretable machine learning: a case study in perspective. Procedia Comput. Sci. **204**, 869–876 (2022). https://doi.org/10.1016/j.procs.2022.08.105

9. Troncoso-García, A.R., Martínez-Ballesteros, M., Martínez-Álvarez, F., Troncoso, A.: Explainable machine learning for sleep apnea prediction. Procedia Comput. Sci. **207**, 2930–2939 (2022). https://doi.org/10.1016/j.procs.2022.09.351

10. Nigar, N., Umar, M., Shahzad, M.K., Islam, S., Abalo, D.: A deep learning approach based on explainable artificial intelligence for skin lesion classification. IEEE Access **10**, 113715–113725 (2022). https://doi.org/10.1109/ACCESS.2022.3217217

11. Civit-Masot, J., Bañuls-Beaterio, A., Domínguez-Morales, M., Rivas-Pérez, M., Muñoz-Saavedra, L., Rodríguez Corral, J.M.: Non-small cell lung cancer diagnosis aid with histopathological images using explainable deep learning techniques. Comput. Methods Programs Biomed. **226**, 107108 (2022). https://doi.org/10.1016/j.cmpb.2022.107108

12. Loh, H.W., et al.: Deep neural network technique for automated detection of ADHD and CD using ECG signal. Comput. Methods Programs Biomed. **241**(107775), 107775 (2023). https://doi.org/10.1016/j.cmpb.2023.107775

13. Zou, L., et al.: Ensemble image explainable AI (XAI) algorithm for severe community-acquired pneumonia and COVID-19 respiratory infections. IEEE Trans. Artif. Intell. **4**(2), 242–254 (2023). https://doi.org/10.1109/TAI.2022.3153754

14. Khater, T., et al.: An explainable artificial intelligence model for the classification of breast cancer. IEEE Access **1** (2023). https://doi.org/10.1109/ACCESS.2023.3308446

15. Amado-Caballero, P., Casaseca-de-la-Higuera, P., Alberola-López, S., Andrés-de-Llano, J.M., López-Villalobos, J.A., Alberola-López, C.: Insight into ADHD diagnosis with deep learning on actimetry: quantitative interpretation of occlusion maps in age and gender sub-groups. Artif. Intell. Med. **143**(102630), 102630 (2023). https://doi.org/10.1016/j.artmed.2023.102630

16. Yilmaz, R., Yagin, F.H., Raza, A., Colak, C., Akinci, T.C.: Assessment of hematological pre-dictors via explainable artificial intelligence in the prediction of acute myocardial infarction. IEEE Access **11**, 108591–108602 (2023). https://doi.org/10.1109/ACCESS.2023.3321509

17. Saravanan, S., Ramkumar, K., Narasimhan, K., Vairavasundaram, S., Kotecha, K., Abraham, A.: Explainable artificial intelligence (EXAI) models for early prediction of Parkinson's disease based on spiral and wave drawings. IEEE Access **11**, 68366–68378 (2023). https://doi.org/10.1109/ACCESS.2023.3291406

18. Camacho, M., et al.: Explainable classification of Parkinson's disease using deep learning trained on a large multi-center database of T1-weighted MRI datasets. NeuroImage Clin. **38**(103405), 103405 (2023). https://doi.org/10.1016/j.nicl.2023.103405

19. Sheu, R.-K., Pardeshi, M.S., Pai, K.-C., Chen, L.-C., Wu, C.-L., Chen, W.-C.: Interpretable classification of pneumonia infection using eXplainable AI (XAI-ICP). IEEE Access **11**, 28896–28919 (2023). https://doi.org/10.1109/ACCESS.2023.3255403

20. Lysdahlgaard, S.: Utilizing heat maps as explainable artificial intelligence for detecting abnor-malities on wrist and elbow radiographs. Radiography (Lond.) **29**(6), 1132–1138 (2023). https://doi.org/10.1016/j.radi.2023.09.012

21. Islam, M.K., Rahman, M.M., Ali, M.S., Mahim, S.M., Miah, M.S.: Enhancing lung abnor-malities detection and classification using a deep convolutional neural network and GRU with explainable AI: a promising approach for accurate diagnosis. Mach. Learn. Appl. **14**(100492), 100492 (2023). https://doi.org/10.1016/j.mlwa.2023.100492

22. Ramírez-Mena, A., Andrés-León, E., Alvarez-Cubero, M.J., Anguita-Ruiz, A., Martinez-Gonzalez, L.J., Alcala-Fdez, J.: Explainable artificial intelligence to predict and identify prostate cancer tissue by gene expression. Comput. Methods Programs Biomed. **240**, 107719 (2023). https://doi.org/10.1016/j.cmpb.2023.107719

23. Bellantuono, L., et al.: An eXplainable artificial intelligence analysis of Raman spectra for thyroid cancer diagnosis. Sci. Rep. **13**(1), 16590 (2023). https://doi.org/10.1038/s41598-023-43856-7

24. Hossain, M.M., et al.: Cardiovascular disease identification using a hybrid CNN-LSTM model with explainable AI. Inform. Med. Unlocked **42**(101370), 101370 (2023). https://doi.org/10.1016/j.imu.2023.101370

25. Moreno-Sánchez, P.A.: Data-driven early diagnosis of chronic kidney disease: development and evaluation of an explainable AI model. IEEE Access **11**, 38359–38369 (2023). https://doi.org/10.1109/ACCESS.2023.3264270

26. Nayak, T., et al.: Deep learning based detection of monkeypox virus using skin lesion images. Med. Nov. Technol. Devices **18**, 100243 (2023). https://doi.org/10.1016/j.medntd.2023.100243

27. Muscato, F., Corti, A., Gambaro, F.M., Chiappetta, K., Loppini, M., Corino, V.D.A.: Combining deep learning and machine learning for the automatic identification of hip prosthesis failure: development, validation and explainability analysis. Int. J. Med. Inf. **176**, 105095 (2023). https://doi.org/10.1016/j.ijmedinf.2023.105095

28. Varam, D., et al.: Wireless capsule endoscopy image classification: an explainable AI approach. IEEE Access **11**, 105262–105280 (2023). https://doi.org/10.1109/ACCESS.2023.3319068

29. Massafra, R., et al.: Analyzing breast cancer invasive disease event classification through explainable artificial intelligence. Front. Med. **10**, 1116354 (2023). https://doi.org/10.3389/fmed.2023.1116354

30. Nkengue, M.J., Zeng, X., Koehl, L., Tao, X.: X-RCRNet: an explainable deep-learning network for COVID-19 detection using ECG beat signals. Biomed. Signal Process. Control **87**(105424), 105424 (2024). https://doi.org/10.1016/j.bspc.2023.105424

31. Jiménez-García, J., et al.: An explainable deep-learning architecture for pediatric sleep apnea identification from overnight airflow and oximetry signals. Biomed. Signal Process. Control **87**(105490), 105490 (2024). https://doi.org/10.1016/j.bspc.2023.105490

32. Cozma, G.V., Onchis, D., Istin, C., Petrache, I.A.: Explainable machine learning solution for observing optimal surgery timings in thoracic cancer diagnosis. Appl. Sci. (Basel) **12**(13), 6506 (2022). https://doi.org/10.3390/app12136506

33. Tao, S., Ravindranath, R., Wang, S.Y.: Predicting glaucoma progression to surgery with artificial intelligence survival models. Ophthalmol. Sci. **3**(4), 100336 (2023). https://doi.org/10.1016/j.xops.2023.100336

34. To, T., et al.: Deep learning classification of deep ultraviolet fluorescence images toward intra-operative margin assessment in breast cancer. Front. Oncol. **13**, 1179025 (2023). https://doi.org/10.3389/fonc.2023.1179025

Author Index

A

Abdallah, Asma Ben   243
Algeelani, Nasir Ahmed   85
Al-Obeidat, Feras   85
Amin, Adnan   85
Anthopoulos, Leonidas   150

B

Balderas-Díaz, Sara   74
Ballon-Bahamondes, Roberto Daniel   199
Barba-Guaman, Luis   54
Bastardo, Rute   107
Bedoui, M. Hedi   243
Behzadi-Khormouji, Hamed   158
Boubeta-Puig, Juan   74
Burbano, Ricardo   97

C

Cárdenas-Rueda, Pedro Ronald   199
Cardoso, Waldson Rodrigues   34
Chaim, Ricardo Matos   170
Chaouch, Aymen   243
Chicaiza, Janneth   54
Conceição, Luís   44
Corrente, Gustavo   44

D

Domínguez, Federico   158
Duarte, M. Salomé   64
Durães, Dalila   64

E

Estévez, Eduardo   97
Euko, Joao Paulo   13

F

Fernandes, Afonso   118
Freitas, Alberto   44

G

Gaspoz, Cédric   24
Gonçalves, Raquel   254
Graça, Manuel   118
Grubisic, Viviane V. F.   170
Grzelak, Sławomir   232
Guerrero-Contreras, Gabriel   74

H

Henriques, Ana   254
Henriques, Inês   129
Hmida, Badii   243

J

Joachimiak, Michał   232
Jusas, Vacius   211

K

Karakidi, Maria   150
Koctong-Choy, Amanda Hilda   199

L

La Cruz, Kevin Mario Laura-De   199
Llerena, Lucrecia   97

M

Marcondes, Francisco S.   64
Marques, Eliane Cunha   170
Marques, Jorge   139
Marreiros, Goreti   44
Martins, Gilberto   64
Martins, Rafael   44
Mesgaribarzi, Niusha   3
Messaoud, Nada Hadj   243
Meszyński, Sebastian   232
Mohamed, Amr F.   211
Monteiro, Simone B. S.   170
Moreira, Fernando   85
Muñoz, Andrés   74

N
Novais, Paulo   13, 64

O
Oliveira, Pedro   64

P
Pacheco, João   139
Pardo, Eddy   54
Parola, Henrique   254
Pavão, João   107
Payen, Laurent   243
Peláez, Enrique   158

Q
Quevedo, Sebastián   158

R
Ribeiro, Admilson de Ribamar Lima   34
Rocha, Álvaro   188
Rocha, Nelson Pacheco   107, 118, 129, 139
Rodrigues, Manuel   254
Rodríguez, Nancy   97
Rosa, Gilberto   129
Rosselet, Ulysse   24

S
Saad, Jamal   243
Santos, Ema   118
Santos, Flavio   13
Santos, Joana   139
Schiller, Marcin   232
Shudaiber, Ahmed   85
Silva, João Marco Cardoso da   34
Sokolov, Oleksandr   232
Sousa, Maria José   188
Sousa, Miguel   188
Souza, Júlio   44
Szkulmowski, Zbigniew   232

T
Tavares, Florbela   129
Tenorio, Alex Eduardo Tapia-   199
Tozo-Burgos, Jose Giancarlo   199
Tselios, Dimitrios   150

V
Valdiviezo-Diaz, Priscila   54
Valle, Luz Anabella Mendoza-Del   199

X
Xavier, William   44

Printed in the United States
by Baker & Taylor Publisher Services